CHEMICAL EVOLUTION

MOLECULAR EVOLUTION TOWARDS THE ORIGIN OF LIVING SYSTEMS ON THE EARTH AND ELSEWHERE

CHEMICAL EVOLUTION

Molecular Evolution Towards the Origin

of Living Systems

on the Earth and Elsewhere

MELVIN CALVIN

1969

OXFORD UNIVERSITY PRESS

New York and Oxford

PREFACE

THE lectures upon which this book is based had their beginnings in 1949 during a period of recuperation after an illness. As for most men, musings on the beginnings of man had long played a part in my thoughts, but during this particular period they became focused. After this, pieces of the puzzle of the beginnings of man and, beyond this, the beginnings of life itself became increasingly open to extended evaluation and experiment. The tools of chemistry and physics were developing in precision, and concepts of biochemistry and biology were evolving in generality, so that investigations into the earliest period—that borderland between the non-living and the living—became possible.

From the earliest of the modern experimental observations, beginning in 1950, has grown a large body of knowledge from laboratory experiments and from observation in the field. A realm of thought that had only recently been largely speculative, or philosophical, has now become one of scientific investigation and the chief occupation of a large number of experimental and observational scientists.

The study of chemical evolution is based upon the assumption that life appeared on the surface of the earth as a result of the normal operation of the laws of physics and chemistry. This implies that there must have been a period of time in the earth's history that encompassed the transition between a non-living molecular population on its surface and a population of molecular aggregates that we would call living. The two approaches used in this book to try to determine the nature of that transition and interface are those of molecular palaeontology—that is, the look backwards in the historical record—and the construction of hypothetical chemical systems that could give rise to living organisms.

This is a very personal essay, based largely upon work done in our laboratories, although, of course, it calls also upon the relevant information that has appeared elsewhere. It is not an exhaustive documentary account of the subject.

For the past half-dozen years my colleagues and I have been discussing, both publicly and privately, various aspects of the problem of the origin of life, and of man, and it had been my growing hope that some opportunity would present itself that would enable me to record in some more durable form the results of these discussions, as well as some observations. This opportunity came during my year as Eastman

Professor at Oxford in 1967–8, and the present volume is based on the lectures and discussions that took place during that period at Oxford. Its production was made possible by the patience, perseverance, and co-operation of many people, beginning with my scientific host at Oxford, Professor R. W. Richards, and his staff. The initial transcription of the recorded discussions was made by Mrs. Marilyn Taylor in Berkeley; this and the editing and documentation of the final copy represent the primary creation of the manuscript. The final creation of the book from a necessarily rough draft is the work of the expert and co-operative staff of the Clarendon Press.

From its conception to its final chapter, much of which she wrote, my wife, Genevieve, is everywhere in this book.

M. C.

Berkeley, California
November 1968

CONTENTS

PART II

THE VIEW FROM THE PAST TOWARDS THE PRESENT

INTRODUCTION

THE subject of this essay, more directly and generally put, might be the major human question. Since man is his own most vital concern, the title might have been 'Man: his past, present, and future'. I therefore have all the scope I need to undertake a series of discussions that could go on indefinitely. There are in fact a number of specific human relevancies in what we are going to discuss, and it will be useful if I explain what these relevancies are for me, and what they might possibly be for others, in the terminology of man's past, present, and future.

By *past*, I mean the study of all of man's past, from the beginning of the earth in its present form, leading to the development of man's view of himself in relation to the whole universe. This would begin perhaps with cosmology and cannot end, at the earliest, before the present moment.

The *present* of this threefold discussion would be the more technical one: how 'things' work, illuminated in part by the search for how things got that way, which is part of the study of evolutionary history. This leads almost immediately to the ability to manipulate our present environment. The ability to manipulate stems, essentially, from our understanding; and it leads to the whole of technology (applied knowledge): biology to medicine and agriculture, chemistry to applied chemistry (the making of new materials), and physics to engineering (the making of new 'machines'). There are a whole series of activities that stem from the increase in our fundamental knowledge of how 'things' work. The application of that knowledge to the environment is the manipulation of our environment.

In terms of man's *future*, there will certainly be insights to be derived into the essential characteristics of living and thinking systems. These insights will have profound influence, not only in the way in which we might conduct our imminent exploration of the solar system, but also on how, and in what direction, we are likely to change the social and political organization of mankind.

Ever since man became conscious of himself, it seems he has been profoundly concerned with his own nature. In fact, the degree to which he has left evidence of this concern is frequently taken as a measure of his progress toward the human condition. Very early in his speculations about his own nature, man recognized that he was a member of a large class of objects on the surface of the earth that were called 'living' as distinguished from those that were not. Very soon, as a corollary, or extension, of his concern about his own nature man became interested in the nature of life and living things themselves. In the course of his history, both ancient and modern, man produced an enormous variety of notions about both the nature of life and his relation to it. One of the first examples of man's expression of the temporal and hierarchical sequence of events by which he arrived at the present moment and condition is in the book of Genesis. This presents a temporal and hierarchical sequence that contains the essential notion of evolution itself: the first day, the light; the second day, the stars; the third day, the earth; the fourth day, the sun and the moon; the fifth day, the fish in the sea and the fowl in the air; the sixth day, man; and the seventh day, rest. This is the germ of the basic idea of a hierarchical sequence in time. In a sense, we shall be concerned in this book with the fifth day, more or less.

There are other ways in which this idea has been represented. Recently there has been a literary expression by Thornton Wilder in his book *The Eighth Day*: 'Nature never sleeps. The process of life never stands still. The creation has not come to an end. The Bible says that God created man on the sixth day and rested, but each one of these days was many millions of years long.' (Wilder continues: 'That day of rest must have been a short one. Man is not an end but a beginning. We are at the beginning of the second week. We are children of the eighth day.')[1] The idea of the hierarchical sequence of evolutionary events has been represented graphically by the modern Dutch artist Maurits Escher, one of whose pictures, 'Verbum' (Fig. 0.1), is an excellent representation of the whole idea of evolutionary history.[2] It is perhaps of some interest to note that as the birds of the night give way to the birds of the day (from left to right in Fig. 0.1) the highest form is not that of the hawk, the eagle, or the falcon but rather the form of the dove. Perhaps one might even discern a small twig in the beak of the most complete one, the one furthest west!

In modern times the scientific expression of evolutionary development takes its form with Darwin. Darwin and Wallace's paper published in 1858 has the title 'On the tendency of the species to form varieties and on the perpetuation of the species by natural means of selection'. The original Wallace title, which never appeared on a publication, is simpler and more graphic: 'On the tendency of varieties to depart indefinitely from original types'.[3] This title (which appeared in the manuscript that Wallace sent to Darwin in 1858) is a most expressive description of the essence of the evolutionary principle, in the way I shall apply it, because accepting that notion and then extrapolating backwards in the diagrammatic way shown in Fig. 0.2 leads to the origin of life from a non-living milieu.

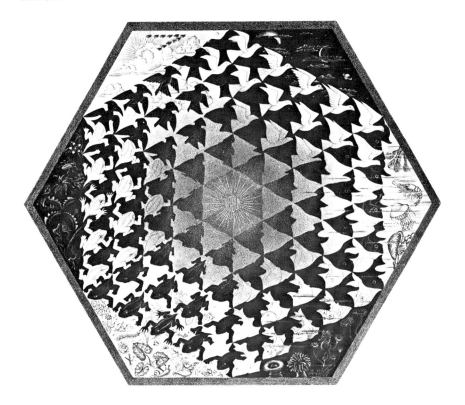

FIG. 0.1. 'Verbum.' Maurits Escher. (Reproduced by permission of the artist.)

Eventually, when we go back far enough in time, we find only a single living species that could, in turn, be conceived of as one of a variety of physico-chemical systems, most of which were not alive, which had been generated from the primeval molecular systems with which the earth began. The interface between the two evolutionary systems is probably a diffuse thing and it seems best to represent it that way.

This really is the basic notion to which this entire book is devoted, and it is perhaps worth while using Darwin's own words to describe the fact that he also recognized that implication of the evolutionary theory. His words, in a letter to Wallich in 1882, are:

You expressed quite correctly my views where you said that I had intentionally left the question of the Origin of Life uncanvassed as being altogether *ultra vires* in the present state of our knowledge, and that I dealt only with the manner of succession. I have met with no evidence that seems in the least trustworthy, in favour of so-called Spontaneous Generation. I believe that I have somewhere said (but cannot find the passage) that the principle of continuity renders it probable that the principle of life will hereafter be shown to be a part, or consequence, of some general law. . . .

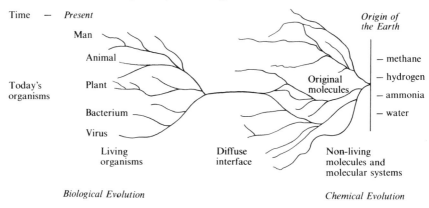

Fig. 0.2. Diagrammatic representation of origin of life from a non-living milieu.

The statement to which Darwin refers, and which he had forgotten, was written earlier, in 1871 (see Fig. 0.3):†

† The complete text of the letter is as follows:
February 1/71

Down,
Beckenham, Kent. S.E.

My dear Hooker,

I return the pamphlets, which I have been very glad to read.—It will be a curious discovery if Mr. Lowne's observation that boiling does not kill certain molds is proved true; but then how on earth is the absence of all living things in Pasteur's experiments to be accounted for?—I am always delighted to see a word in favour of Pangenesis, which some day, I believe, will have a resurrection. Mr. Dyer's paper strikes [?] me as a very able Spencerian production.

It is often said that all the conditions for the first production of a living organism are now present, which could ever have been present. But if (and oh what a big if) we could conceive in some warm little pond with all sorts of ammonia and phosphoric salts,—light, heat, electricity &c. present, that a protein compound was chemically formed, ready to undergo still more complex changes, at the present day such matter w^d be instantly devoured, or absorbed, which would not have been the case before living creatures were formed.

Henrietta makes hardly any progress, and God knows when she will be well.

I enjoyed much the visit of you four gentlemen, i.e., after the Saturday night, when I thought I was quite done for.

Yours affect^y
C. Darwin

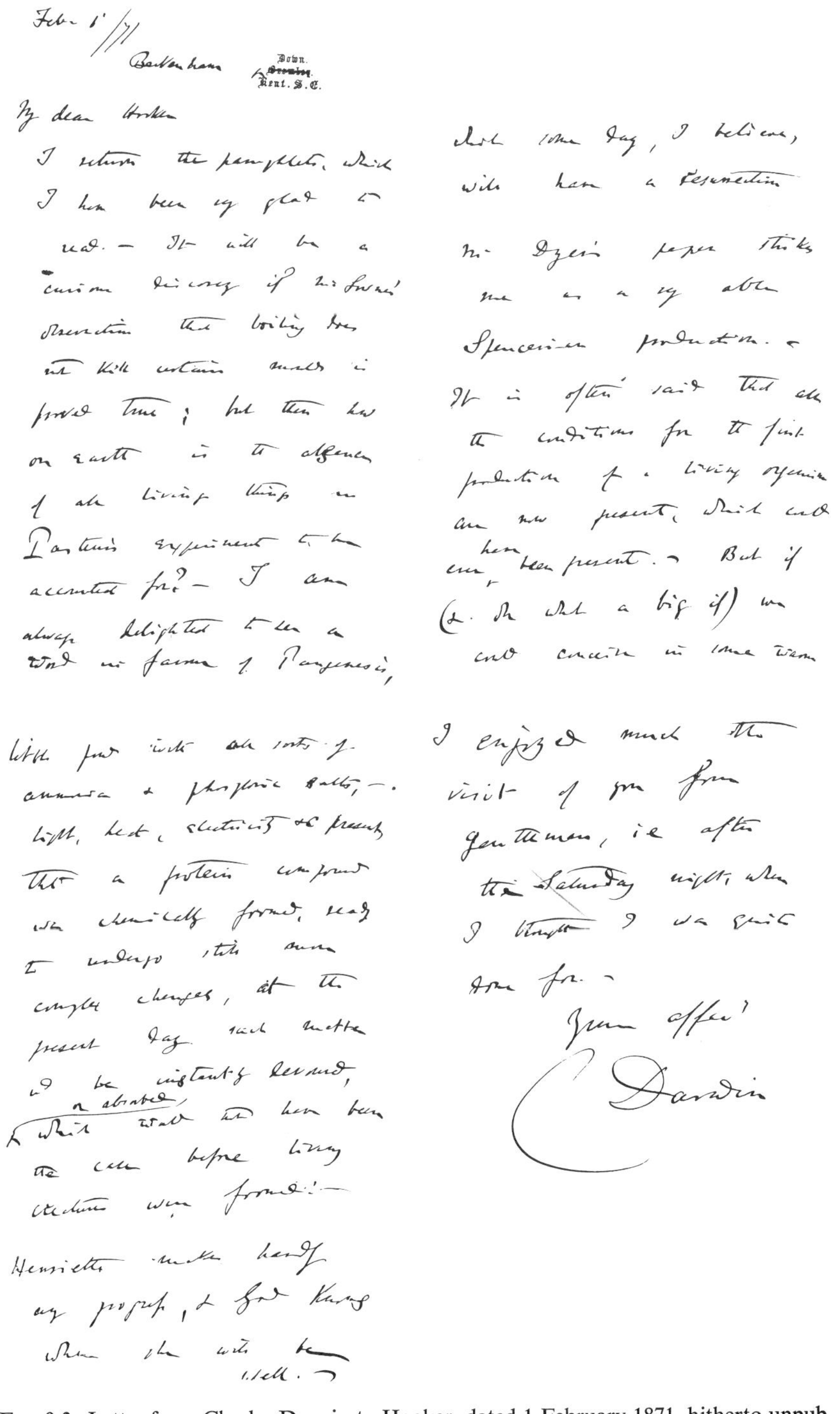

FIG. 0.3. Letter from Charles Darwin to Hooker, dated 1 February 1871, hitherto unpublished. (Reproduced by courtesy of Dr. R. C. Olby and Dr. Peter Gautrey.)

It is often said that all the conditions for the first production of a living organism are now present, which could ever have been present. But if (and oh what a big if) we could conceive in some warm little pond, with all sorts of ammonia and phosphoric acid salts, light, heat, electricity, etc., present, that a protein compound was chemically formed, ready to undergo still more complex changes, at the present day such matter would be instantly devoured, or absorbed, which would not have been the case before living creatures were formed.

Darwin had the idea of 'chemical' evolution, but he was also aware of the limitations of the scientific knowledge of his day; that both the biology and chemistry of his day were not sufficiently well developed to allow him to make much progress along these lines, in speculating about the nature of these primitive events.

The situation is different today from what it was 100 (or 150) years ago. Modern developments both in the knowledge of the fundamental laws of molecular behaviour, molecular interactions, and changes on the one hand, and our better knowledge of the detailed nature of living things on the other hand, make it profitable to undertake this exercise and this study.[4] The plan of organization of the present book in exploring the sources of scientific knowledge about this matter is as follows.

(i) *The view from the present towards the past*

We shall begin with an examination of the historical record, trying to read it from the present time backward into history. This is really a kind of palaeontology, which in the first part of our study is organic geochemistry and finally turns into what I like to call *molecular palaeontology*.

(ii) *The view from the past towards the present*

This will be the view that begins with the primeval earth and moves forward in time toward the present. We shall try to make these two views meet and try to find the diffuse area (interface) between them; and possibly, in the course of the search, define it.

(iii) *The view towards the future from the present*

This view is, perhaps, more fun than the others, partly because most of us can have opinions about it without being concerned about whether those opinions are confirmed or denied. However, I want to include a discussion of the imminent extra-terrestrial exploration we are about to undertake—or have already begun, in our examination of the meteorites and the moon—of Mars and Venus, and the impact that any observation of some sort of organism or organic compound will have on the definition of the qualities of the matter we seek that will guide us in further exploration. The design of the devices and experiments to explore not only the moon, but our two nearest planetary neighbours, Venus and Mars, has already progressed some distance along this road. And it is quite obvious that any observation of an extra-terrestrial physico-chemical system to which the appellation 'living' might be applied will have enormous effects on our concepts of man's place in the universe, comparable to the two prior revolutions that science has imposed on philosophy, namely, the removal of the earth from the centre of the universe and the removal of the need for special creation.

Finally, in this view toward the future these observations of extra-terrestrial systems will help toward the formation of clearer concepts of the essential characteristics of living systems. This ought to enable us to devise more intelligently the changes that will take place in our political and social organization.

(iv) *In search of human significance*

I shall also say a few words about the significance of the entire idea of evolutionary exploration in terms of human activity.

REFERENCES

1. WILDER, THORNTON, *The eighth day*, p. 22. Popular Library edition, New York (1967).
2. *The graphic work of M. C. Escher*. Duell, Sloan & Pierce, New York (1967); Oldbourne, London (1967).
3. DARWIN, C., and WALLACE, A. R. On the tendency of the species to form varieties and on the perpetuation of the species by natural means of selection. *J. Linn. Soc.* (Zoology) **3**, 45 (1858).
   WALLACE, A. R. On the tendency of varieties to depart indefinitely from original types. Unpublished MSS. written in June 1858 and incorporated into the longer Darwin–Wallace paper mentioned above.
4. For reviews of the general theories of the origin of life on the earth, see references below, in addition to the books listed in the bibliography at the end of the volume.
   (*a*) Collection of essays on the origin of life, *New biology*, No. 16. Penguin Books, London (1954).
   (*b*) GAFFRON, H., The origin of life. *Grad. J.* **4**, 82 (1961). See also GAFFRON, H., The origin of life. *Perspect. Biol. Med.* **3**, 163 (1960).
   (*c*) FOX, S. W., How did life begin? *Science* **132**, 200 (1960).
   (*d*) HOROWITZ, N. H., and MILLER, S. L., The origin of life. *Fortschr. Chem. org. NatStoffe* **20**, 423 (1962).
   (*e*) WALD, G., The origins of life. *Proc. natn. Acad. Sci. U.S.A.* **52**, 595 (1964).

(f) CALVIN, M., Chemical evolution (the Bakerian lecture). *Proc. R. Soc.* A **288**, 441 (1965).

(g) PONNAMPERUMA, C., and GABEL, N. W., Current status of chemical studies on the origin of life. *Space Life Sci.* **1**, 64 (1968).

(h) LEMMON, R. M., Abiogenic synthesis of biologically-relevant organic compounds ('chemical evolution'). *University of California Lawrence Radiation Laboratory Report No. UCRL* 18108 (1968); *Chem. Rev.*, in preparation.

(i) A discussion on anomalous aspects of biochemistry of possible significance in discussing the origins and distribution of life. Discussion of The Royal Society, organized by N. W. Pirie, 2 November 1967. *Proc. R. Soc.* B **171**, 2–89 (1968).

PART I

THE VIEW FROM THE PRESENT
TOWARDS THE PAST

1

THE FOSSIL RECORD

FIG. 1.1 is a representation of the geological time-scale, giving some idea of the regions of time, and geology, with which we shall be concerned in our backwards look. The first 600 million years is pretty well documented in classical palaeontology. Going further backward, into the Pre-Cambrian, we shall try to find morphological remains of significance in the region of time earlier than the Cambrian period (600 million-year mark). Beyond the Pre-Cambrian we shall in general run out of morphological remains and we shall then have to depend on 'molecular fossils', molecular remains in which the intimate molecular architecture is the clue, or the indication, of its origin. In beginning this examination it is useful to see the Phanerozoic time-scale, i.e. the first 600 million years in the backward look (Fig. 1.2), which is well described. The division into eras, periods, and epochs has been done mostly by stratigraphical dating and from the fossils themselves as a means of studying the time-scale. A diagrammatic representation of the fossils of the Phanerozoic (the 600-million-year period beginning with the bottom of the Cambrian (Fig. 1.3) gives some idea of how the fossil record has provided information about the appearance and disappearance of a variety of organisms. An enormous variety of organisms appear at the very beginning of the Cambrian period, and basically most of the existing types were present then.[1] It has only been in recent years that the existence of a palaeontological record prior to the Cambrian has been shown. Fig. 1.4 shows more explicitly the relationship between the

Phanerozoic time-scale and the rest of the history of the earth. In the centre of Fig. 1.4 are shown some names of specific rocks and also some events that took place at times corresponding to their position on the time-scale, looking backward from the present. Later we shall examine

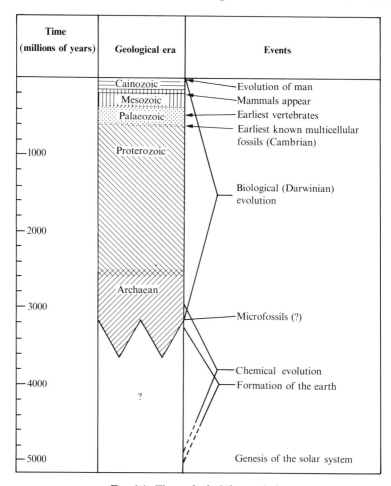

Fig. 1.1. The geological time-scale.†

some of the evidence for the Pre-Cambrian fossil record prior to the 600-million-year mark.

In the last decade or two interest in Pre-Cambrian molecular remains has grown enormously, and there have been a whole series of reports

† The time-scale subdivision for the Pre-Cambrian into late, middle, and early Proterozoic and Archean is now under discussion (Glaessner, private communication, July 1968).

(which really extend for a longer period than the twenty years implied) in which presumed Pre-Cambrian fossils were reported. Fig. 1.5 shows the geographical regions of the earth in which Pre-Cambrian fossils have been reported. They are scattered all over the globe, although the

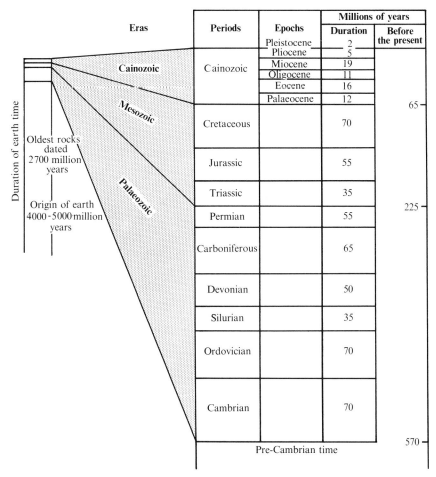

Eras	Periods	Epochs	Millions of years	
			Duration	Before the present
Cainozoic	Cainozoic	Pleistocene	2	
		Pliocene	5	
		Miocene	19	
		Oligocene	11	
		Eocene	16	
		Palaeocene	12	65
Mesozoic	Cretaceous		70	
	Jurassic		55	
	Triassic		35	225
Palaeozoic	Permian		55	
	Carboniferous		65	
	Devonian		50	
	Silurian		35	
	Ordovician		70	
	Cambrian		70	570

Duration of earth time

Oldest rocks dated 2700 million years

Origin of earth 4000-5000 million years

Pre-Cambrian time

FIG. 1.2. The Phanerozoic time-scale.

Canadian Pre-Cambrian shield, the South African Pre-Cambrian shield, and the Australian Pre-Cambrian formations have the majority. The rock-samples I shall discuss are mostly from the Lake Superior region (Michigan), the south-east African region (Swaziland), and central Australia.

At this point it may be worth mentioning the methods that are used for dating the ancient rocks. Most of this dating is now done by isotopic

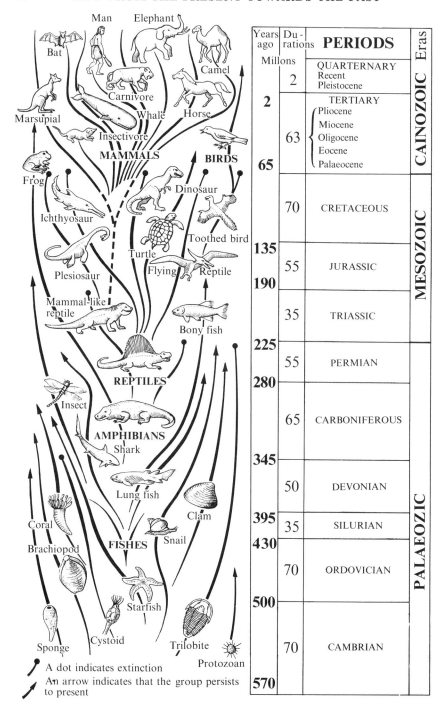

Years ago	Du-rations	PERIODS		Eras
Millons				
	2	QUARTERNARY Recent Pleistocene		CAINOZOIC
2	63	TERTIARY Pliocene Miocene Oligocene Eocene Palaeocene		
65				
	70	CRETACEOUS		MESOZOIC
135	55	JURASSIC		
190	35	TRIASSIC		
225	55	PERMIAN		
280	65	CARBONIFEROUS		PALAEOZIC
345	50	DEVONIAN		
395	35	SILURIAN		
430	70	ORDOVICIAN		
500				
570	70	CAMBRIAN		

FIG. 1.3. The palaeontological record of the Palaeozoic.

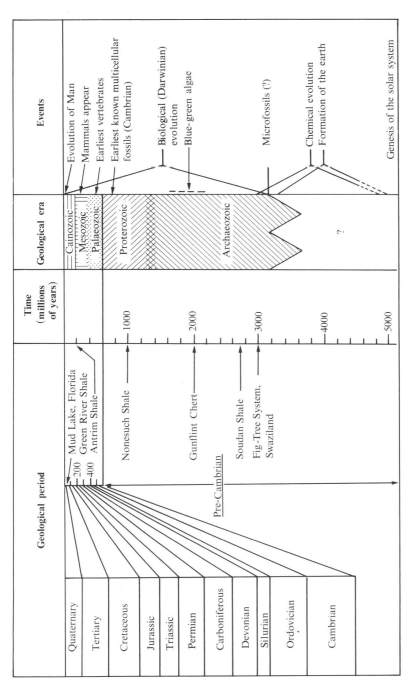

FIG. 1.4. Relationship between the Phanerozoic time-scale and the history of the earth.

methods.[2] These depend upon our ability to measure particular isotopes of certain elements at extremely low concentrations and with high precision. The two particular isotope ratios of importance are the potassium–argon system and the rubidium–strontium system:

$$^{40}K \rightarrow {}^{40}Ar \quad \text{(half-life, } 12 \cdot 4 \times 10^9 \text{ years),}$$
$$^{87}Rb \rightarrow {}^{87}Sr \quad \text{(half-life } 50 \cdot 0 \times 10^9 \text{ years).}$$

Potassium and rubidium are both naturally-occurring radioactive elements, that is, there is a spontaneous transformation from potassium to

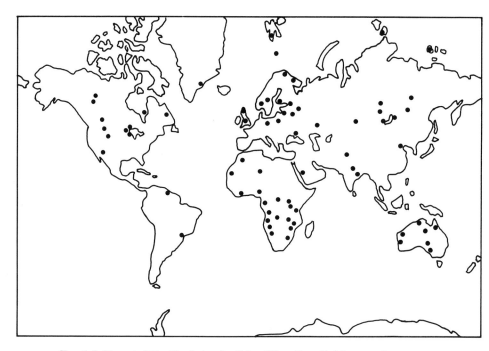

FIG. 1.5. Reported Pre-Cambrian fossil localities. Compiled from various sources by G. E. Murray (1965).

argon and from rubidium to strontium (argon-40 and strontium-87 are stable but potassium-40 and rubidium-87 are not). Each of these transformations is the result of a series of decays, just as in the thorium and uranium series. The thorium and uranium half-lives are similar to those for potassium–rubidium, i.e.

$$^{232}Th \rightarrow {}^{208}Pb \quad \text{(half-life } 13 \cdot 9 \times 10^9 \text{ years),}$$
$$^{235}U \rightarrow {}^{207}Pb \quad \text{(half-life } 0 \cdot 7 \times 10^9 \text{ years),}$$
$$^{238}U \rightarrow {}^{206}Pb \quad \text{(half-life } 4 \cdot 5 \times 10^9 \text{ years).}$$

The way in which these isotopes are used to determine the age of a rock is by measuring the ratio of the amount of potassium-40 to the amount of argon-40 in some particular sample. The assumptions made are that at the time the particular rock took a solid crystalline form there was no argon-40 in it; that after the rock solidified no gas escaped; and that all of the argon-40 formed subsequent to solidification is still in the rock. This type of dating method thus requires a certain kind of rock, one that has cooled from a melt. A sediment may be intruded by a magmatic dyke or some other sort of igneous intrusion; and since the sediment is clearly now cut by the intrusion, we can establish that the sediment is older than the igneous rock. Therefore, by dating the igneous intrusion we can determine the youngest age possible for the sediment into which it has been intruded. This type of dating operation is usually done in more than one geological location with more than one isotope before a date is really established. The dates given in the discussion for the sedimentary rocks are based upon isotope dating methods, and the minimum age of the sediment can be accepted as being fairly accurately determined.

The first set of Pre-Cambrian sediments to be discussed here are the youngest—remember this is a view toward the past—in the late Pre-Cambrian, roughly 650 million years old. In these rocks, in spite of the discontinuity between the Cambrian and Pre-Cambrian rocks, a whole set of fossils have been documented relatively recently. There has been recognized for a long time a type of structure that has been called a 'fossil', which is presumably a residue formed as a result of the action of a living organism. These are the stromatolitic formations in many of the ancient Pre-Cambrian rocks, going back as much as 1500 million years, and reaching up to the present.[3] They are called 'fossils' because these layered, bent formations of limestone are today visible in the living or just recently dead blue-green algae one finds in very quiet, or very slow, waters. The blue-green algae grow and settle down on top of each other in mats, and the calcareous residues of such algal mats, called *stromatolites*, are well known. In addition to these, in the late Pre-Cambrian a number of higher or more complex animal forms have been described. Fig. 1.6 shows a segmented worm from rocks about 650 million years old in the Ediocara Hills in Central Australia, which shows a clear structure of some sort.[4,5] This fossil was discovered by Glaessner, who usually shows beside it a photograph of a modern animal of similar appearance. Some blue-green algae, dated at 800 million years, from Bitter Springs, Australia, are shown in Fig. 1.7.[6] These show very

FIG. 1.6. Segmented worm, *Spriggina floundersi*, from Pre-Cambrian of Central Australia, which resembles certain present-day forms of segmented worms. (From Glaessner.)

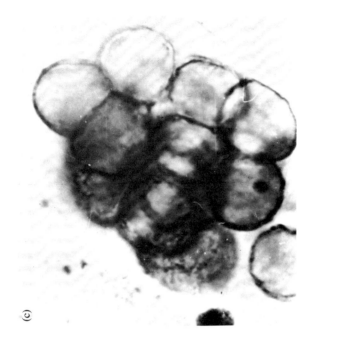

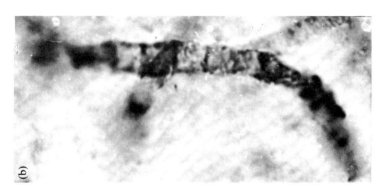

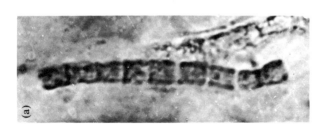

Fig. 1.7. Organisms from Bitter Springs Chert, Australia (7 to 9×10^9 years old). (From Barghoorn and Schopf 1965.)

clearly a segmented form similar to certain blue-green algae of today. These are relatively highly developed organisms.

Pls. 1–3 illustrate some of the rocks to be mentioned later. At California we have used the Green River Shale (Pl. 1), which is about 60 million years old, as a 'model' substance, because it is so rich in hydrocarbons and fossils; it is used as a 'testing ground' for our various analytical methods. Pl. 2 shows the Nonesuch Shale (containing oil) from North America. Pl. 3 shows the Soudan Chert, a hard black rock containing about 2 per cent carbon. The Nonesuch, the Gunflint, and the Soudan rocks are ones from which have been isolated a variety of organic chemicals, whose structure will be discussed later. In some of these rocks a variety of microfossils have been described, particularly of the algal type. Fig. 1.8 shows the microfossils found in the Gunflint Chert. This again shows, although much older (1700 million years) than the Australian rock (800 million years), similar filamentous blue-green algae.[7,8] Fig. 1.9 illustrates a polished section of the Gunflint Chert etched with hydrofluoric acid, showing the organic coated residues, presumed to be bacteria (the organic coated residues have the shapes of small coliform types of bacteria, a few micrometres in length). Keep in mind the age of this sample at 1700 million years. Similar fossil-like bacteria have also been found in much older rocks. Fig. 1.10 shows another object found in the Gunflint section. At the time this material was published someone who was working for NASA, and had been looking at micro-organisms from all over the world in strange environments, thought he recognized this particular shape. He therefore hunted through his records to find a micro-organism he thought was the same, with the cap shape and a stem. He found one: it was cultured from the soil around the base of the walls of Harlech Castle, in Wales (Fig. 1.11). This has a particular significance, because it turns out that this organism, found at Harlech Castle, requires an environment of 30 per cent ammonia to grow. My point in mentioning this is that the original atmosphere of the earth is supposed to have been rich in ammonia, and the resemblance is so striking, and the requirements for ammonia so suggestive, that I could not resist the comparison. How does it happen that one finds these ammonia-loving organisms at the base of Harlech Castle? Mr. H. P. Powell, assistant to the curator of geological collections at Oxford, comments that Harlech Castle 'had been there for 700–800 years and the base of the walls have been enriched with urea for a long time, and is still in effect producing an enriched medium for the selection of ammonia-requiring organisms'. If there is any such

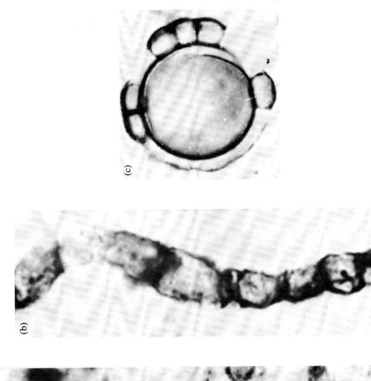

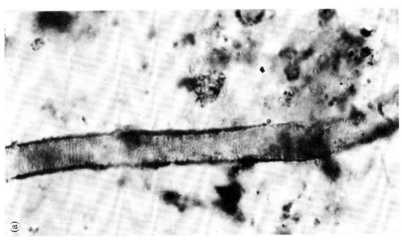

FIG. 1.8. Microfossils from Gunflint Chert. (From Barghoorn and Tyler 1965.)

Fig. 1.9. Etched polished section of Gunflint Chert ($1 \cdot 7 \times 10^9$ years old), showing presumed organic bacterial residues.

significance to these facts then a search for these ammonia-requiring organisms in other parts of the world might be expected to reveal them widespread in suitable environments.

We have now found a very high degree of organization as far back as 1700 million years, and all the microfossils we have seen at this particular period are micro-organisms. We have passed the time when any macro-fossils, that is, things that would have the degree of organization of the Pre-Cambrian segmented worm shown in Fig. 1.6, are so far found.

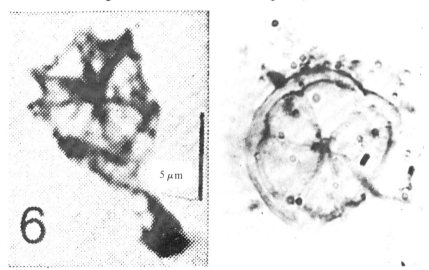

FIG. 1.10. Microfossil from Gunflint Chert (left) compared with cultured specimen from soil at Harlech Castle (right).

The last rock in this series is the Fig Tree Chert from South Africa; dated at 3100 million years, it is far and away the oldest. A sample of the Fig Tree Chert is shown in Pl. 4. Because the rock is so hard it is easy to study a section of the rock after it has been cut. The bacterial fossils found in the Fig Tree Chert are shown in Fig. 1.12, which shows the crystal boundaries and polishing scratches running right through the remnants of the centres of the objects, proving that the objects were already present when the chert crystallized.[9] I want to point to the possi-bility that we are dealing with organisms 3100 million years of age; and if these are indeed bacteria that have anything like the complexity of modern bacteria in their metabolic systems, we have, as early as 3100 million years ago, a highly evolved complex metabolic system. A com-parison of the fossil algae in the Fig Tree, Gunflint, and Bitter Springs rocks is shown in Fig. 1.13; (1), (2), (3), and (4) are organisms from the

Fig Tree Chert (3100 million years), (5) and (6) are from Gunflint (1700 million years), and (7) is the blue-green alga from Bitter Springs (800 million years).[10] There is definitely a resemblance among the organisms. The youngest (Bitter Springs) has the most form. The dark grains inside

FIG. 1.11. Harlech Castle, Wales.

the organisms are apparently organic; they do not dissolve in hydrofluoric acid when the rock is etched. This is the kind of argument with which we have to deal and these are materials of the kind that Pre-Cambrian palaeontology is considering. The significance of the 'formed elements' in both the meteorites and the ancient rocks has been called into question.[11,12] Particularly effective in stimulating doubt about the interpretation of the electron micrographs of sections have been the stereoscan electron-microscope pictures of a Pre-Cambrian thucholite[13] and of the Fig Tree Chert itself.[14] It is evident that the structures as seen in section might be some purely physico-chemical curiosity in the formation of the chert itself.

Fig. 1.14 summarizes the fossil history of the rocks.[15] The younger rocks have many different materials in them, and as the rocks get older —the Bitter Springs, the Nonesuch, the Siyeh limestone, the Gunflint— fewer and fewer organisms appear as microfossils. In the Bulawayan

FIG. 1.12. Electron micrographs of *Eobacterium isolatum* in surface replicas of Fig Tree Chert (3×10^9 years old). (a) Organically preserved cell (white) and its imprint, transgressing a polishing scratch in the rock surface. (b) Well-preserved rod-shaped cell. (c) Transverse section of the fossil organism showing the preserved cell wall. (From Barghoorn and Schopf, 1966.)

limestone (of which I have no photograph) some microfossils have been reported, and in the Fig Tree Chert there were only two presumed organisms, the bacteria and the blue-green alga.

We have gone far back in time: all the way back to 3100 million years. The next stage in our investigation will be to determine whether we can confirm these morphological microfossils as true biological residues by some other method, a chemical method. If these are truly biological residues, there should be some molecular fossils with them: some of the

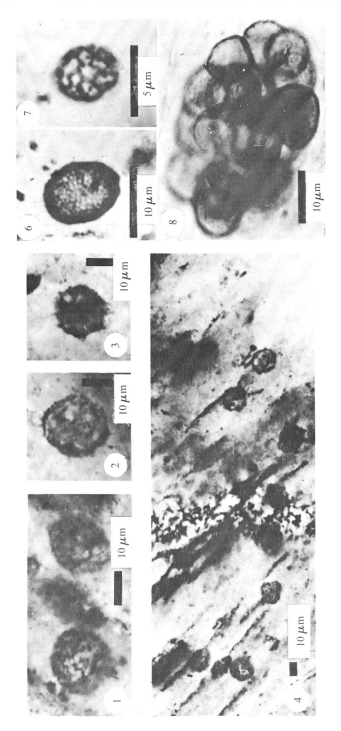

Fig. 1.13. Microfossils from Fig Tree Chert (1–4), Gunflint Chert (5–6), and Bitter Springs Chert (7) compared. (From Schopf and Barghoorn 1967.)

molecules that these microfossils made might be expected to be found in the rocks themselves. With the advanced methods of chemical analysis available today we can find extremely minute quantities of certain kinds

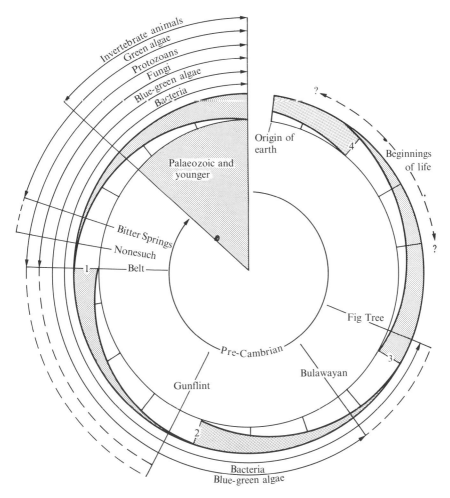

FIG. 1.14. 'Geological clock' showing fossiliferous Pre-Cambrian sediments and time-ranges of primitive organisms. Broken lines indicate uncertain occurrences. Time is shown in units of 10^9 years before the present. (After Schopf.)

of organic materials and can also describe the intimate molecular architecture. From that architecture it should be possible to devise a self-contained scheme that will allow us to know when chemical evolution came to an end and organic evolution began.

REFERENCES

1. STIRTON, R. A., *Time, life and man: The fossil record*. Wiley, New York (1959).
2. For references on isotope dating methods, see:
    (a) CRAIG, H., MILLER, S. L., and WASSERBURG, G. J., *Isotopic and cosmic chemistry*. North-Holland, Amsterdam (1964).
    (b) RANKAMA, K., *Progress in isotope geology*. Wiley, New York (1963).
    (c) POLANSKI, A., *Geochemistry of isotopes*. Waraaw (1965).
    (d) SCHAEFFER, O. A. and ZÄHRINGER, J. (editors) *Potassium-argon dating*. Springer-Verlag, Heidelberg (1966).
    (e) Geochronology of rock systems. *Ann. N.Y. Acad. Sci.* **91**, 159 (1961).
3. GLAESSNER, M. F., Precambrian paleontology. *Earth Sci. Rev.* **1**, 29 (1966).
4. —— Precambrian animals. *Scient. Am.* **204**, No. 3, 72 (1961).
5. —— and WADE, M., The Late Precambrian fossils from Edicara, South Australia. *Palaeontology* **9**, 599 (1966).
6. BARGHOORN, E. S. and SCHOPF, J. W., Microorganisms from the Late Precambrian of Central Australia. *Science* **150**, 337 (1965).
7. —— and TYLER, S. A., Microorganisms from the Gunflint Chert. Ibid. **147**, 563 (1965).
    Additional information on the microorganisms in the Gunflint Chert can be found in:
    (a) CLOUD, P. E., JR. and HAGEN, H., Electron microscopy of the Gunflint microflora: preliminary results. *Proc. natn. Acad. Sci. U.S.A.* **54**, 1 (1965).
    (b) LICARI, G. R. and CLOUD, P. E., JR., Reproductive structures and taxonomic affinities of some nannofossils from the Gunflint Iron formation. Ibid. **59**, 1053 (1968).
8. BARGHOORN, E. S., MEINSCHEIN, W. G., and SCHOPF, J. W., Paleobiology of a Precambrian Shale. *Science* **148**, 461 (1965).
9. —— and SCHOPF, J. W., Microorganisms 3 billion years old from the Precambrian of South Africa (Fig Tree). Ibid. **152**, 758 (1966).
10. SCHOPF, J. W. and BARGHOORN, E. S., Alga-like fossils from the Early Precambrian of South Africa. Ibid. **158**, 673 (1967).
11. BRAMLETTE, M. N., Primitive microfossils or not? Ibid. **158**, 637 (1967).
12. VAN LANGINHAM, S. L., SUN, C. N., and TAN, W. C., Origin of round-body structures in the Orgueil meteorite. *Nature, Lond.* **216**, 252 (1967).
13. PRASHNOWSKY, A. A. and SCHIDLOWSKI, M., Investigation of Precambrian thucholite. Ibid. **216**, 560 (1967).
14. NAGY, B., University of California, San Diego, private communication (1967). For some of the microstructures see NAGY, B., Carbonaceous meteorites. *Endeavour* **27**, 81 (1968).
15. SCHOPF, J. W., Antiquity and evolution of Precambrian life. *McGraw-Hill yearbook of science and technology*, p. 47 (1967).

2

THE STARTING MATERIALS OF
MOLECULAR PALAEONTOLOGY

A ROYAL SOCIETY discussion held in 1967 on 'Anomalous aspects of biochemistry of possible significance in discussing the origins and distribution of life' is relevant to the problem with which we are concerned.[1] The meeting was arranged by Mr. N. W. Pirie, who began by showing on the screen one of the microfossil pictures that we discussed in Chapter 1. He said that this was the kind of evidence we were depending upon for clues as to the earliest data when there were living things on the surface of the earth, and he pointed to the fact that these were purely morphological remains; they were shapes in the rock. Pirie then showed a second slide, which was a picture of a rock formation somewhere near the top of a mountain. He pointed out that there were two physical shapes silhouetted against the sky, as an analogy of the types of images we saw, in principle, in the photographs of the presumed microfossils. One of them was alive (a crouching man); the other was a very interesting shape—and, if I blurred it with my eyes, I thought I could see the outline of President Kennedy's head in it. The moral, as I understood it (although Pirie was a little more subtle about it) was that shapes alone are not enough evidence for us to date the emergence of life on the earth.

In order to find more evidence for the emergence of life, we have chosen to examine the detailed architecture of the organic molecules that we may find associated with morphological shapes, or even in the absence of these shapes, in the rock. Is it possible to deduce from the intimate architecture of the molecules we find in the rocks something about the nature of their origin? Is there some aspect of the structure of the molecules, intrinsic to that structure, that will enable us to say whether their origin was biological or abiological, whether they are the precursors of living things or the residues of living things?†

One of the subdivisions of this discussion as listed in Chapter 1 was 'The view from the present toward the past'. We took this view in terms of the morphological remains in the ancient rocks. Now we shall try to determine its appearance in terms of the architectural distribution of the molecules we may find.

Fig. 2.1 gives a geological time-scale with slightly more information than we have seen before. It shows the various types of rock-samples, the time involved, the molecular structures we are seeking or have found, and the evolutionary events that occurred, as related to these molecules and the time-scale.

Examination of recent sediments as models for evolution of molecular fossils

The first series of analytical results presented here is for a sediment in the process of formation.[9] Can we recognize in this the beginnings of some of the ancient sediments? What kind of materials are laid down, how long do they last, and how long may we expect them to last? This is the kind of information we are going to look for in sediments that are in the process of formation. The one we chose, the Mud Lake in Florida, had been described by Dr. W. H. Bradley of the U.S. Geological Survey as a freshwater sediment in the process of formation. Dr. Bradley helped us to obtain samples from the bottom of the lake upon which we could perform a molecular analysis. Age-determination had already been done by him. The lake is rather stationary; the water seeps out of the bottom and there is no river inflow. The surface is coated with algae of various kinds. The lake is not very deep and the bottom can be reached quite readily. We thus have mud samples from various depths in the sand at the bottom of the lake. Table 2.1 summarizes the elemental analyses of the algal ooze at the mud–water interface and at various levels below it, as well as the ages of the sediments, determined by carbon-14 dating techniques. Fig. 2.2 gives more information about the physico-chemical parameters at various depths of the Mud Lake. For example, the carbonate level rises as the depth increases, and the free-sulphur level rises also; but the reduction potential falls, as does the pH. The oxygen content of the water must decrease with increasing depth, and this is

† Allied to the search for the precursor molecules in the ancient rocks is similar work, under way in various laboratories, on the elucidation of the genetic relationships, if any, between the organic matter present in meteorites and that in ancient sediments. Some pertinent references are 2–8.

reflected in the liberation of carbon dioxide into the mud. The remaining organic material in the Mud Lake is poor in oxygen. The essential features of this system are loss of oxygen from the compounds in the algal ooze, liberation of carbon dioxide, and accumulation of hydrogen,

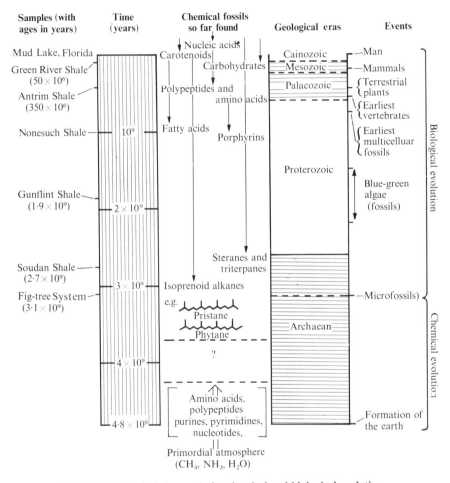

FIG. 2.1. Geological time-scale for chemical and biological evolution.

all of which point to an increase in reducing conditions with age, and thus with depth. This is an important notion, because almost all the current ideas of the diagenetic evolutionary pathways in the rocks depend upon the idea of reduction. Fig. 2.3 shows the isolation and extraction scheme for the materials from the Mud Lake. Three different molecular groups were sought. In the heptane eluate we found hydrocarbons (compounds made up only of carbon and hydrogen) and

TABLE 2.1

Analyses and ^{14}C ages of sediments in the Mud Lake in Florida

Elemental analyses

%	MW-0	MW-2	MW-3	MW-6	Oil shale
C	40·06	40·36	41·17	52·20	80·50
H	6·39	6·45	6·87	5·32	10·30
O	—	—	—	—	5·75
N	4·16	3·82	—	1·39	2·39
S	1·13	2·62	—	2·59	1·04
Ash	—	—	—	11	—

^{14}C ages of sediments

MW-1	1900±200 years
MW-3	2280±250 years
MW-4	4100±250 years
MW-6	5200±250 years

MW-0, Mud Lake algal ooze at mud–water interface.
MW-1–6, Mud Lake algal ooze at 1–6 ft, respectively,
    below the mud–water interface.

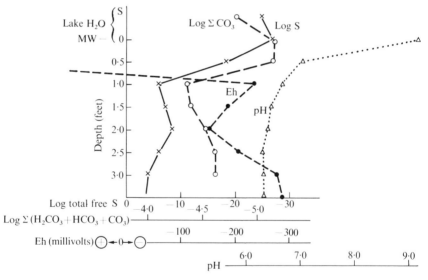

FIG. 2.2. Physico-chemical parameters of sediments from various depths in the
Mud Lake, Florida.

carotenoids, and in the methanol eluate pheophytin a, representative of
the porphyrin group of compounds.

Let us briefly examine the identification procedure for these three
groups of compounds: the porphyrins, the carotenoids, and the hydrocarbons. Fig. 2.4 is a comparison of the absorption spectra of standard
pheophytin with that of the intermediate depth sample, MW-3, from

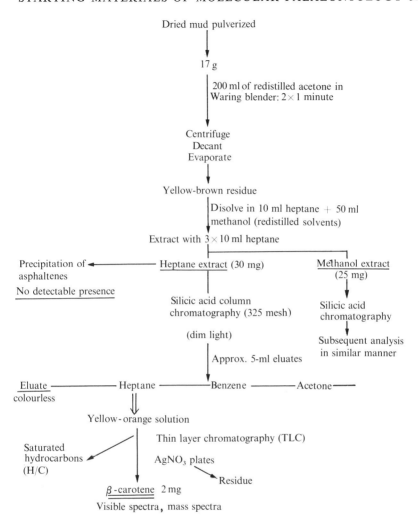

FIG. 2.3. Isolation and extraction scheme for materials from the Mud Lake, Florida.

the Mud Lake; the spectra of the two materials are obviously very similar. Fig. 2.5 shows the relationship between chlorophyll a, and the origin of the pheophytin (chlorophyll a from which the magnesium has been removed), which eventually goes on to become the petroporphyrin. The vanadyl etioporphyrin, shown in the lower part of Fig. 2.5, is the kind of porphyrin found in petroleum. It is related to the chlorophyll molecule by having the side chains reduced, and the magnesium replaced by a vanadyl ion. This shows the first transformation of this kind of material that took place in the mud at Mud Lake, Florida. Table 2.2

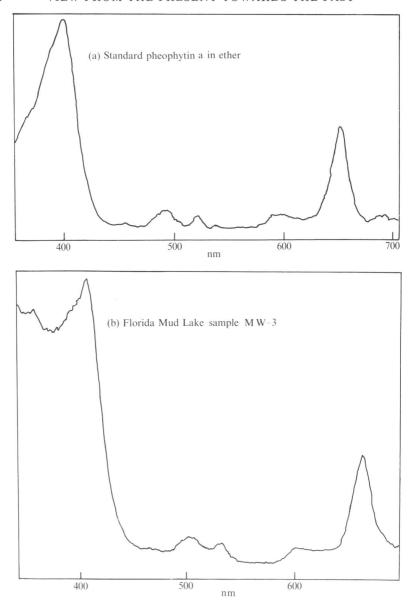

FIG. 2.4. Comparison of absorption spectra of (a) standard pheophytin, and (b) sample from Mud Lake, Florida.

shows carotenoids identified in the Mud Lake sediments. β-carotene (free of oxygen) is very easy to discern. Other carotenoids present have retained their oxygen content even up to 2000 years. Table 2.3 shows carotenoids found in still older rocks up to 100 000 years in age.

Let us now move on to the next variety of molecular 'fossils', i.e. the hydrocarbons that have been found in the Mud Lake oozes approximately 2000 years old. Fig. 2.6 shows a gas chromatogram of the total

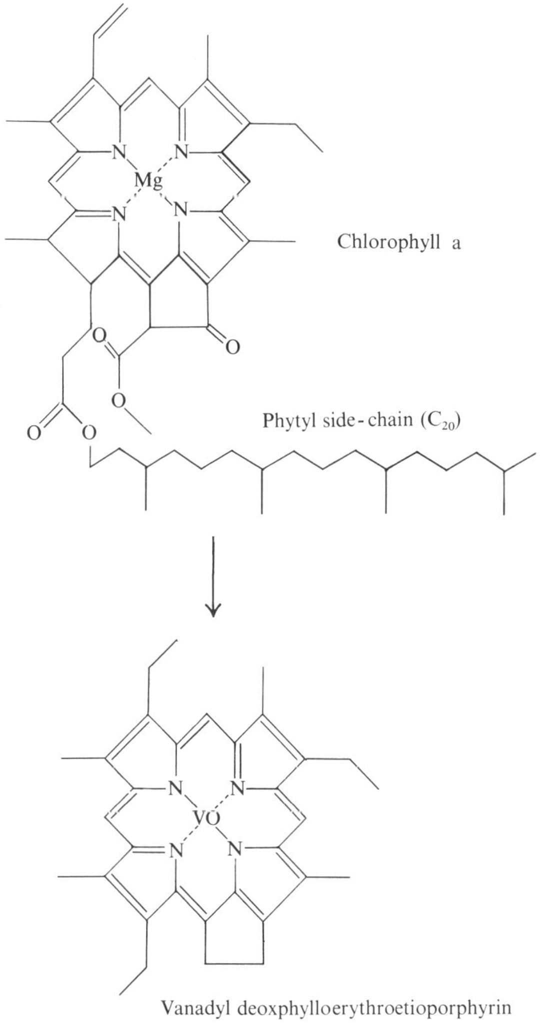

Chlorophyll a

Phytyl side-chain (C_{20})

Vanadyl deoxyphylloerythroetioporphyrin

FIG. 2.5. Relationship between chlorophyll and the petroporphyrins.

alkanes found in the Florida Mud Lake sediments. Gas chromatography is an analytical method that can be used on microgramme quantities of unknown material. The method depends basically upon amounts of compound passing through the apparatus at a rate that is a function of their volatility; the most volatile come through first and the least volatile last. The compounds come out individually, one after the other. The separation pattern is dominated by the normal hydrocarbons, C_{20} to C_{31}, that is, the unbranched carbon chains are the principal ones. Note also that in the Mud Lake alkane extract *all* the normal hydrocarbons from C_{17} upwards appear—there are no missing ones. There is a certain amount of odd–even alternation, but in general the odd-numbered hydrocarbons dominate, C_{20} and C_{22} being rather high and C_{21} rather low.

TABLE 2.2. *Carotenoids from Mud Lake, Florida*

β-carotene

Mass spectra
m/e 536 (M)

Visible spectra (in heptane as solvent)

Mud Lake	Standard
(nm)	(nm)
476	479
466	469
449	451
428–9	430

The maximum of the n-hydrocarbon distribution seems to be at about C_{25} to C_{30}. This is to be kept in mind for comparison later with what is found in living micro-organisms.

Lifetimes of molecular 'fossil' types as potential residues

I should like to discuss the stabilities of these classes of molecular 'fossils'. There are two important classes of materials that I have not yet mentioned, namely amino acids coming from the peptides, and carbohydrates coming from various kinds of polysaccharides. I have not described the amino acids or carbohydrates of the mud because both these compounds may be expected to, and do, disappear quite rapidly. They do not remain as stable compounds for very long periods of time in any large amounts.

The relative chemical and thermal stability of the carbohydrates, amino acids, carotenoids, porphyrins, and hydrocarbons increases in that order. It is very easy to char, or caramelize, carbohydrates (sugars, cellulose), and the lifetime of material of this type is much shorter even than that of the amino acids. One does not, therefore, find sugar present in sediments for very long. In fact, several distribution profiles as a function of the age of the sediment have been made, and both carbohydrates and amino acids disappear quite rapidly. A direct amino acid stability test was performed by Philip Abelson in 1959, during the course of his examination of fossil shells, to try to estimate how long the amino acid would survive. Some carbon-14-labelled amino acids were prepared, labelled both in the carboxyl group and in other parts of the molecule, and it was possible to measure decomposition by ^{14}C evolution from the heated amino acid. The rate of decomposition was extremely small. Abelson calculated the activation energy ΔH^* for carbon-14-labelled alanine, for which A was 3×10^{13} in the equation $k = A \exp(-\Delta H^*/RT)$. ΔH^* was 44 000 calories. k is a measure of the rate of disappearance of the amino acid with time, an ordinary unimolecular disappearance. The time required to reduce the initial amino acid to a value of $1/e$ of its initial amount at room temperature (300 °K) is about 10^9 years. If

TABLE 2.3. *Carotenoids from sediments*

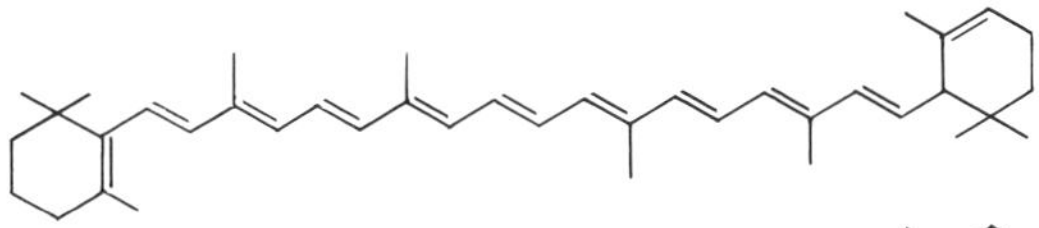

α-carotene

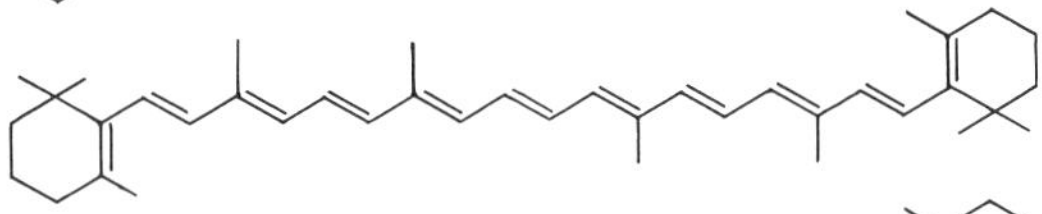

β-carotene

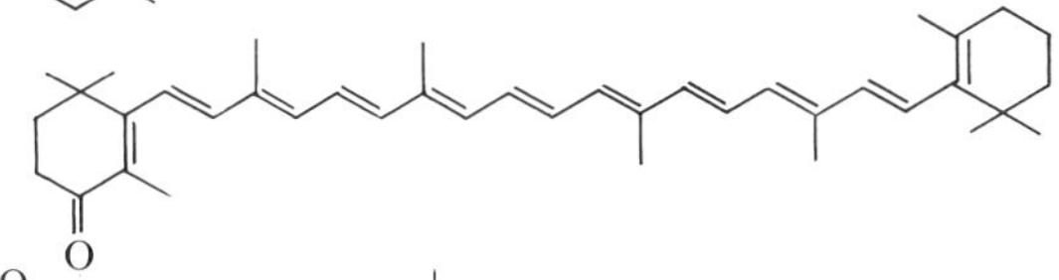

Echinenone

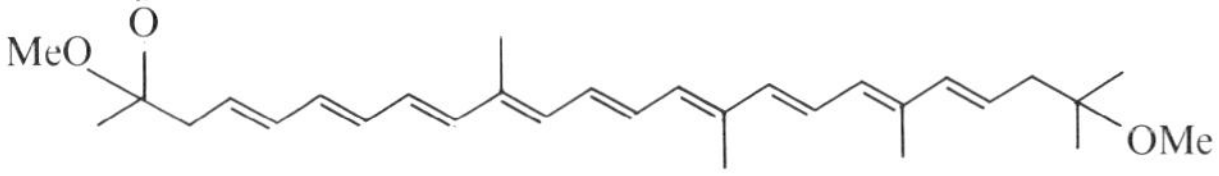

Rhodoviolascin or spirilloxanthin

(1) Oldest sediment (Gytta deposit)   100 000 years
  β-carotene   (Andersen and Gundersen, 1955)[10]

(2) Recent sediments (*a*) 20 000 years		(Vallentyne 1957)[11]
	β-carotene Neo β-carotene U (*cis*-isomer) Echinenone	
	(*b*) 11 000 years	(Vallentyne 1956)[12]
	α-, β-carotenes Echinenone Rhodoviolascin Lycopene	
	(*c*) 8000 years	(Fox 1944)[13]
	Carotenes Xanthophylls	

Xanthophyll

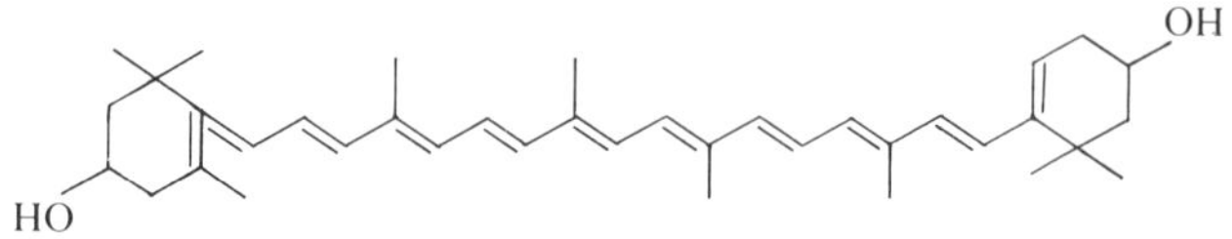

Visible spectra (in benzene as solvent)

Mud Lake (nm)	Standard (nm)
482	487
473	482
455	461
431	437

Intense red coloration in CS_2

Infra-red

—OH group; 3300 cm^{-1}

Rhodoxanthin

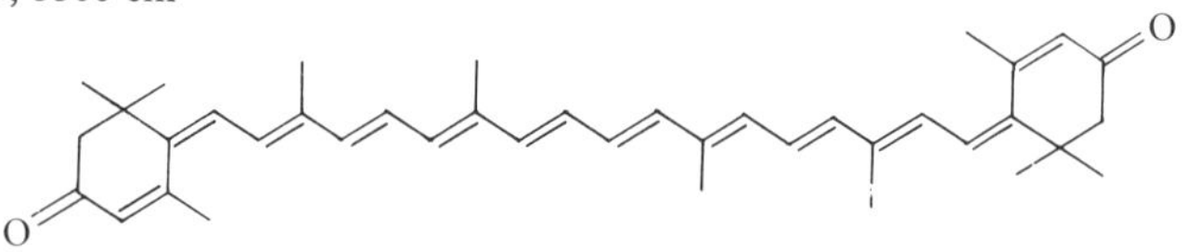

Visible spectra

Hexane (nm)	Benzene (nm)	Cyclohexane (nm)	Methanol (nm)
461	478	464	456
484	500	490	481
517	531	512	512

the lifetime is determined for 400 °K it turns out to be 10^3 years. This tells us that we cannot expect very much amino acid to survive for any great length of time. This is an ordinary unimolecular disappearance, with no allowance made for catalytic effects or complexing stabilization. It is clear that amino acids will not survive long enough in the sediments for our purposes. What about the other types of molecules? The carbohydrates, of course, will disappear even more rapidly, in a few minutes at 400 °K. One sediment profile gave a reduction to 10^{-4} of the original level in 10 years, including both thermal and microbial decay.[14]

The porphyrin ring will probably last considerably longer than the amino acids, for it is a more stable structure.[15,16] The porphyrins can be sublimed at 300 °C (600 °K). The only direct observational information I have about the porphyrin stability is that the porphyrin: hydrocarbon ratio in a recent sediment is less than 1:10 and in a young crude oil the ratio has already been reduced to 1:10 000. This suggests that the porphyrin is less stable than the hydrocarbon by a factor of 1000.

For the hydrocarbons, however, we do have experimental evidence, because the business of cracking hydrocarbons has been going on for quite a long time in industry. For a formal unimolecular rate

$$k = A \exp(-\Delta H^*/RT),$$

the value of A is of the order of 10^{14}, with ΔH^* 66 500 for the breaking of a carbon–carbon bond. Using these figures we can determine the time required for the amount of this compound to decrease to $1/2\cdot3$ times its original value. At 300 °K it would be 10^{27} years, and at 400 °K it would be $10^{14\cdot5}$ years.

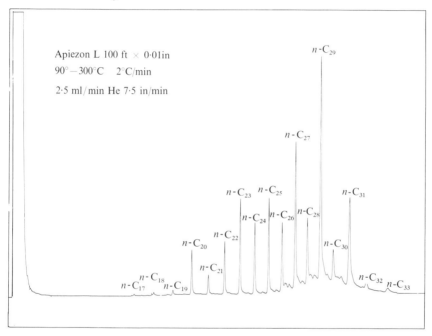

Fig. 2.6. Gas chromatogram of total alkanes in Florida Mud Lake sediments.

This is not the only way in which hydrocarbons can be destroyed or removed. They can be destroyed by breaking off a pair of hydrogen atoms (in fact this is the way most industrial cracking is done). For this process the value of A is about 10^{13} and ΔH^* 63 000 cal. Using these figures, we can determine the lifetimes of the hydrocarbons as 10^{25} years at 300 °K and $10^{13\cdot5}$ years at 400 °K.

Thus, the hydrocarbons are the most stable group of compounds and may be expected to retain a significant part of their original molecular architecture. And we may expect the structural patterns to be helpful in deducing the origin of the hydrocarbons in the oldest rocks. What we want to do eventually is to be able to discern the interface between abiological precursors, or substrates, of life and biological residues of life. We want to be able to recognize that period in time when the abiological organic material was present simultaneously with that produced by a living thing. Only from the internal intimate molecular architecture of these hydrocarbons, and their distribution, can this information be derived.[17] The hydrocarbons seem to be the only group that is likely to have survived long enough to reach that period of time, which we now think is beyond the 2000-million-year mark and perhaps even 3500 million years.

REFERENCES

1. A discussion on anomalous aspects of biochemistry of possible significance in discussing the origins and distribution of life, organized by N. W. Pirie, November 1967. *Proc. R. Soc.* B171, 2–89 (1968).
2. NAGY, B., Carbonaceous meteorites. *Endeavour* 27, 81 (1968).
3. HAYES, J. M., Organic constituents of meteorites—a review. *Geochim. cosmochim. Acta* 31, 1395 (1967).
4. OLSON, R. J., ORÓ, J., and ZLATKIS, A., Organic compounds in meteorites. II. Aromatic hydrocarbons. Ibid. 31, 1935 (1967).
5. COMMINS, B. T. and HARRINGTON, J. S., Polycyclic aromatic hydrocarbons in carbonaceous meteorites. *Nature, Lond.* 212, 273 (1966).
6. DEGENS, E. T., Genetic relationships between the organic matter in meteorites and sediments. Ibid. 201, 1092 (1964).
7. HODGSON, G. W. and BAKER, B. L., Evidence for porphyrins in the Orgueil meteorite. Ibid. 202, 125 (1964).
8. BRIGGS, M. H. and MAMIKUNIAN, G., Organic constituents of the carbonaceous chondrites. *Space Sci. Rev.* 1, 647 (1962–3).
9. HAN, J., McCARTHY, E. D., VAN HOEVEN, W., CALVIN, M., and BRADLEY, W. H., Organic geochemical studies. II. The distribution of aliphatic hydrocarbons in algae, bacteria and in a recent lake sediment: a preliminary report. *Proc. natn Acad. Sci. U.S.A.* 59, 29 (1968).
10. ANDERSEN, S. TH. and GUNDERSEN, K., Ether soluble pigments in interglacial gyttja. *Experientia* 11, 345 (1955).
11. VALLENTYNE, J. R., Carotenoids in a 20 000-year-old sediment from Searles Lake, California. *Archs Biochem. Biophys.* 70, 29 (1957).
12. —— Epiphasic carotenoids in post-glacial lake sediments. *Limnol. Oceanogr.* 1, 252 (1956).
13. FOX, D. L., Biochemical fossils. *Science* 100, 111 (1944).
14. ABELSON, P. H., Paleobiochemistry and organic geochemistry. *Fortschr. Chem. org. NatStoffe* 17, 379 (1959).
15. HODGSON, G. W., BAKER, B. L., and PEAKE, E., Geochemistry of porphyrins. *Fundamental aspects of petroleum geochemistry* (editors B. NAGY and U. COLOMBO), pp. 177–259. Elsevier, Amsterdam (1967).
16. HODGSON, G. W., HITCHON, B., TAGUCHI, K., BAKER, B. L., and PEAKE, E., Geochemistry of porphyrins, chlorins and polycyclic aromatics in solids, sediments and sedimentary rocks. *Geochim. cosmochim. Acta* 32, 737 (1968).
17. For a more detailed discussion, see McCARTHY, E. D. and CALVIN, M. Organic geochemical studies. I. The molecular criteria for hydrocarbon genesis. *Nature, Lond.* 216, 642 (1967).

3

THE NATURE OF HYDROCARBONS
IN MICRO-ORGANISMS

IN addition to learning the nature of the hydrocarbons that might have existed in micro-organisms thousands of millions of years ago, we also need to know more exactly the nature of the material we are starting with today, so that we can make some estimates of what to expect as we go back in time. This immediately leads to a comparative examination of the hydrocarbon skeletons that can be found in toda ̣ 's micro-organisms. I shall therefore describe the comparative examination of a variety of micro-organisms, bacteria, algae, and yeasts for their hydrocarbon and fatty acid content.[1]

The microfossils seen in the ancient rocks are two or three different bacteria, algae, and fungi, particularly the first two types of micro-organisms.

The methods used for analysis of complex material of this type in microgramme amounts are shown in Fig. 3.1. After preliminary extraction of the material in a suitable solvent the resulting material is analysed by gas chromatography, which gives peaks such as those shown in Fig. 2.6. The first approach to identification is simply to know where certain kinds of compounds come on these chromatographic columns, and later to mix the known with the unknown materials to determine identity. However, this is not enough to solve the problem. In addition we now use mass spectrometry, which is an analytical method for finding out first the absolute weight of the molecule, after it has been made into a positive ion by bombardment with 70-volt electrons. Then the way in which the molecule breaks up under the impact of this large amount of energy can be analysed by passing the positive ion fragments through a magnetic field (and sometimes an electric field as well) to see what the mass of each of the fragments is. By knowing the rules about how the molecules behave on fragmentation as a function of the molecular architecture, we can reconstruct the original arrangement of the atoms in fine detail. The beauty of the mass-spectrometric method of analysis is that it can be done with very minute quantities of the unknown compound, and it is extremely sensitive. One of the disadvantages of the method, however, is that results are obtained, perhaps more than occasionally, that have absolutely no relevance to the material being identified. The initial preparation of the sample that is inserted into the

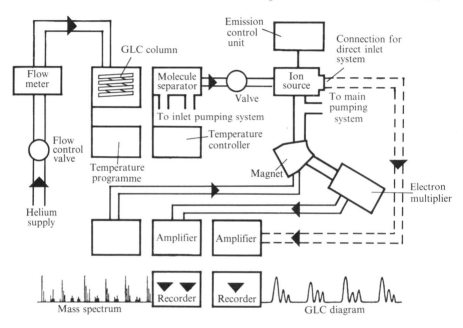

FIG. 3.1. Block diagram of gas chromatograph–mass spectrometer.

G.C.–M.S. analytical chain is of great importance, and care must be exercised from the very beginning in the handling of material for this analysis. This is one of the reasons for what may seem to be incongrously rigorous conditions that are gradually being evolved for the handling not only of the returned lunar samples but of the ancient terrestrial rocks as well. The two methods, gas chromatography and mass spectrometry, are *both* required for the determination of the detailed molecular architecture of the compounds whose structure we need to know if we are to understand the abiological-biological interface.

We shall now return to the discussion of the hydrocarbon content of various micro-organisms,[2] beginning with the analysis of *Chlorella*, an ordinary freshwater green alga with which we have been working in the laboratory.

Fig. 3.2 shows the first steps in the general extraction and fractionation scheme used on *Chlorella*. The same procedure was used on all the other micro-organisms considered here: *Anacystis*, *Rhodospirillum rubrum*, and *Spirogyra* as well as all the others. The hydrocarbon and fatty acid content of the various micro-organisms that we have used are

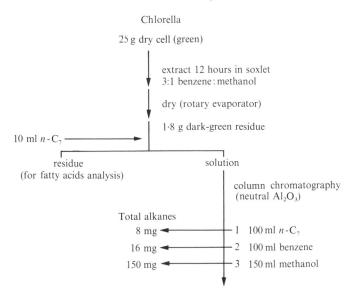

FIG. 3.2. First steps in extraction and fractionation scheme for *Chlorella*.

shown in Table 3.1. Fig. 3.3 shows the total hydrocarbon content of *Chlorella*, as obtained in the n-heptane fraction of Fig. 3.2, and the very large peak at n-C_{17} and Δ-$C_{17}H_{34}$ is clear.

Fig. 3.4 shows the mass spectrum of the n-C_{17} from *Chlorella*. It shows a typical fragmentation pattern characteristic of a normal, unbranched straight-chain hydrocarbon in which the intensity of each successively smaller fragment (differing by one CH_2 group) gets larger, reaching a maximum at about C_4. Fig. 3.5 shows the mass spectrum of the mono-olefinic hydrocarbon from *Chlorella*, Δ-$C_{17}H_{34}$, showing that when the material contains one double bond it is much easier to ionize. The parent molecule minus its electron from the double bond has a considerably higher intensity than any of its neighbours. When there are two pi-electrons, as in a double bond between carbon atoms, it is very easy to eliminate one of them; this ion with one electron missing from the double bond is therefore relatively stable. The general trend is for the smaller ions to be more abundant in much the same regular way as in the normal saturated alkane; this indicates that here also there are no branches, and that the double bond is very near the end of the chain, probably terminal.

TABLE 3.1

Hydrocarbon and fatty-acid content of micro-organisms

	Source	Total alkanes (%)	Fatty acids (%)
Nostoc	Grown in LCB†	0·035	0·025
Anacystis	Grown in LCB†	0·032	0·02
Spirogyra	Bradley, U.S. Dept. of the Interior. Mud Lake, Florida	0·004	0·04
Chlorella	Grown in LCB†	0·032	0·12
Rhodopseudomonas spheroides	Grown in LCB†	0·006	0·16
Rhodospirillum rubrum	Grown in LCB†	0·005	0·2
Micrococcus lysodeikticus	Bought from Miles Laboratories	0·075	0·0015
E. coli	Bought from Miles Laboratories	0·0035	0·9
Yeast	Bought from Co-op. Market	0·005	0·2

The percentages are based on the weight of dry cell.
† Laboratory of Chemical Biodynamics, Berkeley.

We must now continue our comparative biochemical examination of a number of micro-organisms so that we can gain some idea of the kinds of material we might expect to find in the ancient rocks, and ultimately recognize (or seek) in their intimate architecture some clue to the nature of their origin. We have just examined one set of hydrocarbons in the simple micro-organisms, namely the normal hydrocarbons; those molecules in which the carbon atoms are linked together in a continuous chain without branches or rings.

We now examine the gas chromatogram of the total hydrocarbon fraction from another alga, *Spirogyra* (Fig. 3.6). On it appear not only the normal hydrocarbon chains ranging from 16 carbon atoms up to 21 carbon atoms (there are some higher ones that have not yet been identified) but also two other compounds, pristane and phytane. These have not straight hydrocarbon chains but branched ones. This feature is recognized, to a first approximation, by its chromatographic retention time and is finally identified by mass spectrometry. Fig. 3.7 shows the mass spectrum of the *Spirogyra* phytane. This spectrum is interpreted as belonging to a molecule having a chain of 20 carbon atoms with a

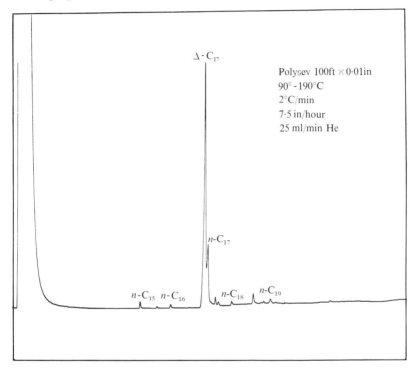

Δ-C_{17}

Polysev 100ft × 0·01in
90° - 190°C
2°C/min
7·5 in/hour
25 ml/min He

n-C_{17}

n-C_{15} n-C_{16} n-C_{18} n-C_{19}

FIG. 3.3. Total hydrocarbon content of *Chlorella* (n-heptane fraction of Fig. 3.2).

branch at every fifth carbon atom, beginning at the second. Interpretation of the mass spectrum is quite straightforward in a case of this kind; the breaks tend to come at the branch points, which are marked on the structural formula below.

We can thus see what kind of fragments can be recognized in that particular mass spectrum.

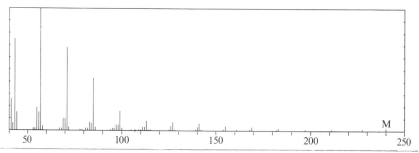

FIG. 3.4. Mass spectrum of n-C_{17} from *Chlorella*.

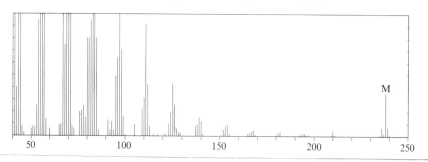

FIG. 3.5. Mass spectrum of mono-olefinic hydrocarbon Δ-$C_{17}H_{34}$ from *Chlorella*, showing ease of ionization of molecule containing a double bond.

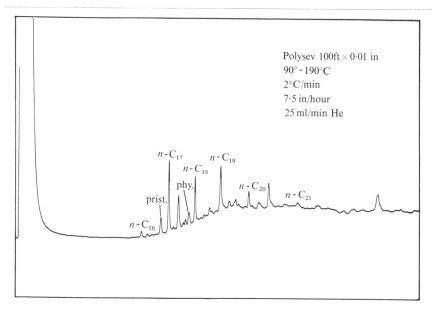

FIG. 3.6. Gas chromatogram of total alkane fraction of *Spirogyra*.

The dominant masses are 197, 183, 127, and 113, which are marked. They come at the break points; the ones near the centre are the most usual.

The other special hydrocarbon marked in the *Spirogyra* chromatogram is the C_{19} compound pristane, whose mass spectrum is shown in Fig. 3.8. The pattern is quite different; 183 is still outstanding, and is

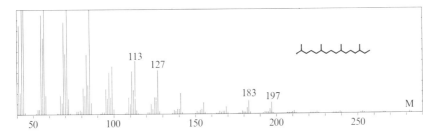

FIG. 3.7. Mass spectrum of phytane from *Spirogyra*.

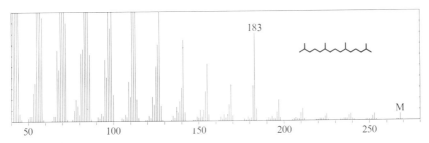

FIG. 3.8. Gas chromatogram of pristane from *Spirogyra*.

the first major peak, with 113 and 127 present as well. The symmetry of pristane is different from that of phytane. The breaks in the centre of pristane are identical, as they are not in the case of phytane; this can be seen from its structural formula:

In the case of phytane, the right-hand half is larger than the left-hand half; in pristane the right-hand and left-hand halves are the same, in terms of mass of fragments.

Pristane and phytane are peculiar compounds. They belong to a large class of compounds called *isoprenoids*, made from the basic unit isoprene, a system of five carbon atoms having a branch at the second one.

$$
\begin{array}{c}
\text{C} \\
| \\
\cdots\text{C--C--C--C--}
\end{array}
$$

The isoprene unit is the element of structure that is common to all these molecules, and it is present in many natural products—terpenes, steroids, carotenoids, and rubber—both as hydrocarbons, in the form shown above, and as a variety of derivatives of hydrocarbons. For example, the carotenoids, which are present in the very young sediments of Mud Lake (Florida), are members of the isoprenoid class.

The significance of the isoprenoids for the questions we have posed should be discussed. These are clearly not completely random structures; they have a branch at every fourth carbon atom. The isoprenoids have a characteristic pattern that is recognizable, particularly by mass spectrometry. We shall endeavour to find out how the micro-organisms construct the isoprenoid-type molecules that are the characteristic skeletons of many of the hydrocarbons found in the ancient rocks.

The specificity of the structure in the biologically produced isoprenoids, and of all their relatives, is known to arise by virtue of a long series of specific enzymatic reactions, the details of which have been worked out by Cornforth[3] and Popjak,[4,5] and Bloch[6] and Lynen.[7] The sequence is well established. There are ten to fifteen individual steps if we begin with the two-carbon fragment, acetate (as shown in the diagrams below). Acetate, which is also one of the first products of the oxidation of ethyl alcohol and gives rise to all the fats (lipids), is transformed, using some of the oxidative energy, to give a molecule in which a thiol ester is formed. The route to the straight-chain unbranched hydrocarbons is the simple one of acetates adding one to another, which necessarily results in an even number of carbon atoms in the molecule, terminating in an acid group. This is why the straight-chain even-numbered acids are the dominant ones, in general, in living organisms. There is another way in which the third acetate could add; it could be added to the middle of the four-carbon keto-acid to a carbonyl instead of to the end. When this is done the result is a branched structure; we can thus see how the branching begins to appear. The branch created is, in effect, the essence of the isoprenoid structure, which is required in order to build up pristane and all the related molecules.

To summarize: one route from the acetate leads to the straight-chain unbranched hydrocarbons; another route from the acetate leads to the five-carbon branched system, which goes on to all the polyisoprenoids with which we shall deal. We can begin with acetate obtainable from CO_2 (or sugar).

$$CH_3 \cdot CO_2H \xrightarrow[\boxed{ATP}]{\boxed{CoA \cdot SH}} CH_3 \cdot CO \cdot S \cdot CoA$$
(acetyl coenzyme A)

1

$$CH_3 \cdot CO \cdot S \cdot CoA \xrightarrow[\boxed{CH_3 \cdot CO \cdot S \cdot CoA}]{\boxed{enzyme \cdot SH}} CH_3 \cdot CO \cdot CH_2 \cdot CO \cdot S \cdot CoA$$
(acetoacetyl coenzyme A)

2

The addition of acetate residues continues linearly to give normal fatty acids with an even number of carbon atoms. Alternatively, the third acetate may condense with the keto group of the acetoacetyl coenzyme A to give branched-chain structures (as shown on following page). Both pristane and phytane are present in modern organisms, such as algae.

This sequence to phytane, for example, involves some ten or more individual steps, each one catalysed by a specific enzyme both in terms of the way in which the materials are hooked together, and the geometry that the resultant molecules achieve. This is a very carefully arranged sequence. What appeared to us as a high degree of specificity, both in the production of the linear chain and in the particular branching system, made us think that this might be an important clue to biological sources. This line of thought is only partly a result of knowledge of the specific enzyme reactions used in present-day living organisms to produce these rather special carbon skeletons, both the straight and branched chains.

The odd hydrocarbons (presumed to come from the even acid by loss of a carboxyl group) and the regularly branched hydrocarbons, the isoprenoids, seem in themselves indications of biological origin. On the other hand, the apparent absence from modern organisms of the very large number of other almost equally possible carbon skeletons that one would expect is significant. To give an idea of what such a mixture of all sorts of things would look like, if we should find them and analyse them, Fig. 3.9 shows the hydrocarbon fraction obtained when iron carbide is dissolved in hydrochloric acid, and only the saturated hydrocarbon fraction is examined.[8] The chromatogram does not separate into distinct recognizable entities. There are so many different compounds

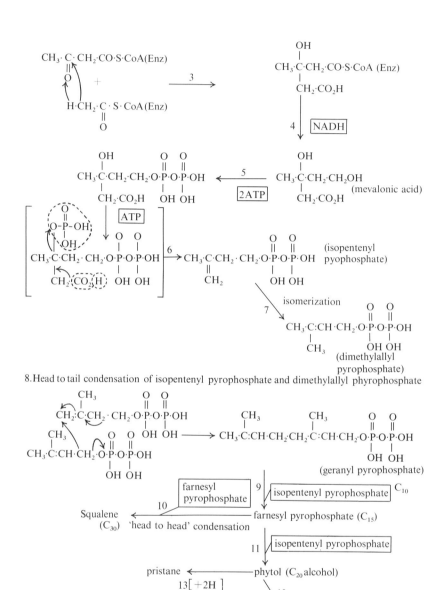

8. Head to tail condensation of isopentenyl pyrophosphate and dimethylallyl phyrophosphate

present that the gas chromatogram simply cannot separate them and they pile on top of each other. There must be thousands of different isomers in this great curve. Perhaps one day we shall find a suitable technique for separating all the isomers (although I do not know exactly why anyone would want to do it). The smoothness of the distribution is evidence that everything in that carbon fraction is clearly abiogenic, which is what one might expect to see if one were to find somewhere deep in the earth a collection of hydrocarbons that were not formed by biological methods. Thus we have some indication of the complexity of

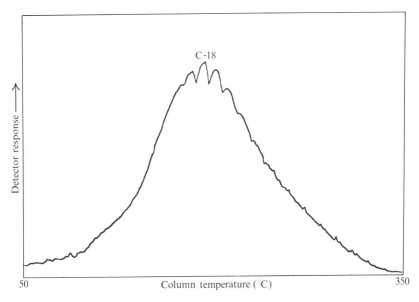

FIG. 3.9. Gas chromatogram for hydrocarbon fraction produced by action of hydrochloric acid on iron carbide. (After Hoering 1966.)

what we might expect in abiogenic hydrocarbons. The fact that we do not find such complex mixtures has been taken as evidence that any given sample was not formed abiogenically. We shall discuss the problem of abiogenic versus biogenic origin again later on when we discuss the mechanisms by which carbon compounds were constructed before the advent of living things.

We now continue our examination of some of the modern organisms for other classes or types of compounds that could be recognized and used as markers in ancient rocks (and lunar rocks also). Fig. 3.10 shows the gas chromatogram of the alkanes from *Nostoc*, a blue-green alga apparently related to the fossils illustrated in Chapter 1. One of the filamentous fossils in the Soudan rock was claimed by Barghoorn to be a microfossil of a blue-green alga, resembling *Nostoc* in form. The chromatogram is heavily dominated by the normal C_{17} hydrocarbon, and there is also the appearance of a *new* branched C_{18}. This branched

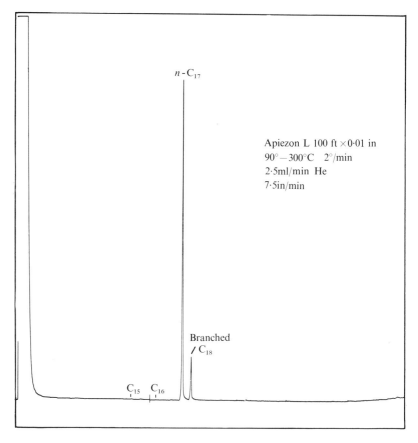

FIG. 3.10. Gas chromatogram for alkanes from the blue-green alga *Nostoc*.

C_{18} is unique, as shown by the mass spectrum (Fig. 3.11),[9] which exhibits a strange variety and pattern of peaks. The first major fragment peak is at 169; the next is at 155. The important feature is that there are two major peaks at 169 and 155, and there are two more at 127 and 113. Recognizing those two major and two subsidiary peaks, it becomes possible to construct a proposed structure for the C_{18} branched compound. The structure is a very peculiar one. The mass of 254 corresponds to 18 carbon atoms, which for a single compound would indicate that there are two branches, separated by one carbon atom, as follows:

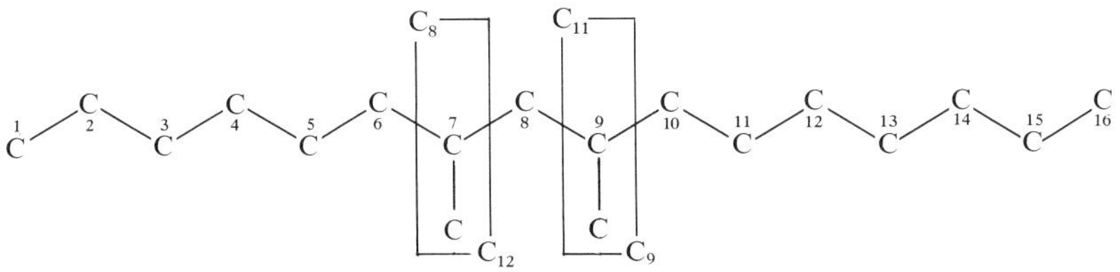

Structure proposed, but later found erroneous, for branched C_{18} compound from *Nostoc*.

The break points should be on the branched carbon atoms in this proposed structure. This would give a straightforward pattern, and perhaps there could be branching at other points; but if that were, in fact, the case the mass spectrum would be slightly different, although this was not certain until the compound itself, or one just like it, could be synthesized. When, in fact, this was done, neither of the two diastereomers synthesized coincided precisely with the *Nostoc* branched C_{18} either chromatographically or mass spectrometrically.[9] This dilemma could be resolved if the *Nostoc* hydrocarbon were supposed to be an unresolved equimolar mixture of 7-methylheptadecane and 8-methylheptadecane:

Synthesis of these two compounds confirmed that supposition in every way, including the dominance of the even-mass fragments of C_8, C_9, C_{11}, and C_{12}.[10] This type of structure determination is an interesting diversion from the comparative biochemistry. I expect that other such anomalies will appear, and eventually we shall have to try to understand how that particular compound or mixture was evolved and constructed. This particular C_{18} branched compound from the *Nostoc* is not part of the regular series—not part of the straight unbranched chain or of the isoprenoid chain. It is something else that may turn out to be part of another regular system about which we know little, and may be unique. If so, it could very well be an important compound for our geochemical search. We have just begun that study, and it continues.

Let us return to a more systematic examination of a few other organisms and groups. Fig. 3.12 shows the hydrocarbons from *Rhodopseudomonas spheroides*, a purple bacterium. Here one can see the same groups of hydrocarbons, dominated by the n-C_{17} compound, with the presence of phytane and pristane as well. This is the latest available of our series on the photosynthetic organisms.

I should now like to examine some non-photosynthetic micro-organisms for their hydrocarbon content to see how they compare with these photosynthetic organisms.[1] A difference in pattern between the photosynthetic and non-photosynthetic organisms here begins to become

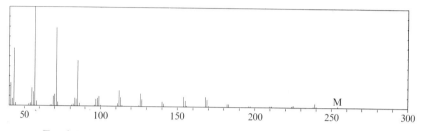

FIG. 3.11. Mass spectrogram of branched C_{18} compound from *Nostoc*.

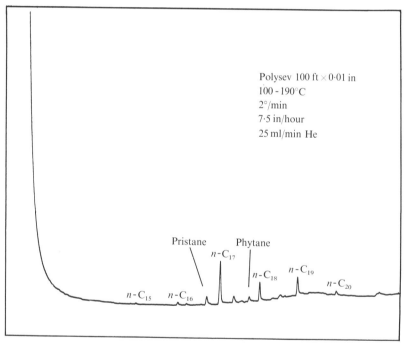

Polysev 100 ft × 0·01 in
100 - 190°C
2°/min
7·5 in/hour
25 ml/min He

Pristane

Phytane

$n\text{-}C_{17}$

$n\text{-}C_{18}$

$n\text{-}C_{19}$

$n\text{-}C_{20}$

$n\text{-}C_{15}$ $n\text{-}C_{16}$

FIG. 3.12. Gas chromatogram of hydrocarbons from *Rhodopseudomonas spheroides*

apparent. Fig. 3.13 shows the total alkane content of *Escherichia coli*, the ordinary *E. coli* that the bacteriologists use, and the difference is becoming apparent. In the non-photosynthetic bacteria the gas chromatogram is no longer dominated by the C_{17} but by the C_{18} compounds; but what is still more important is that there are a large number of

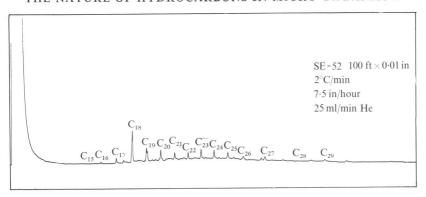

FIG. 3.13. Gas chromatogram of total alkanes from *Escherichia coli* (late log phase).

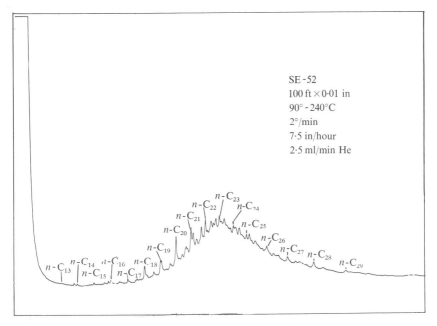

FIG. 3.14. Gas chromatogram of alkanes from yeast.

hydrocarbons of higher molecular weight, around C_{24}, C_{25}, and C_{26} and still higher, that were not apparent in the photosynthetic bacteria as a major component of their hydrocarbon fraction. A striking example of this difference is shown in Fig. 3.14, which is a gas chromatogram of the alkanes from bakers' yeast. It exhibits a curious kind of distribution, in which most of the peaks are in the C_{23} region. Because this shows such a relatively smooth curve it is quite clear that there are many compounds present that our chromatography has not yet succeeded in resolving.

The main point here is that the hydrocarbon fraction of the yeast is of such high molecular weight, in contrast to the non-photosynthetic hydrocarbons. One obligate anaerobe has a gas-chromatographic analysis very much like that of *E. coli*, although its hydrocarbon constitution is in fact slightly heavier than that of *E. coli*. Tables 3.2 and 3.3

TABLE 3.2

Hydrocarbons from algae. Peak heights from mass spectrograms

	Blue-green algae		Green algae	
	Nostoc	*Anacystis*	*Spirogyra*	*Chlorella*
n-C_{15}	0·42	28	—	0·7
n-C_{16}	0·42	3·4	6·7	0·4
Pristane	—	—	22	—
ΔC_{17}	—	—	—	450
n-C_{17}	100	100	100	100
Branched C_{18}	19·4	0·44	—	—
Phytane	—	—	15·5	—
n-C_{18}	0·5	—	58	0·3
n-C_{19}	0·4	—	62	0·1
n-C_{20}	0·4	—	22	—
Alkanes of higher mol. wt.	—	—	Less than 30% of total hydrocarbons	—
Major component	n-C_{17}	n-C_{17}	n-C_{17}	ΔC_{17}

Peak heights are relative to n-C_{17} peak taken as 100.

summarize the hydrocarbons from the algae, bacteria, and yeast, showing that the photosynthetic organisms are dominated by the light hydrocarbon fractions, while the non-photosynthetic organisms are dominated by the heavier hydrocarbon fraction.[2]

I should here like to refer to the hydrocarbon pattern in the Mud Lake sediment (which is only 5000 years old); the lighter hydrocarbons are not dominant, as they are in the fresh photosynthetic organisms. This leads to the general notion that the hydrocarbon content of these relatively young Mud Lake sediments is in fact not the original hydrocarbon directly formed by the photosynthetic organisms but rather a hydrocarbon that has been formed by non-photosynthetic organisms (probably anaerobic ones) at the bottom of the lake, which simply consumed the carbohydrate that the photosynthetic organisms provided as they sank to the bottom. Thus, the hydrocarbons that we find in the Mud

Lake sediments, for example, are more likely to be the products (as a first approximation) of a fermentative transformation of the primary photosynthetic material than the primary photosynthetic material itself. This seems most probable on the simplest consideration. Most of the primary photosynthetic product is actually carbohydrate, and this falls

TABLE 3.3

Hydrocarbons from bacteria and yeast. Peak heights from mass spectrograms

	Photosynthetic bacteria		Non-photosynthetic bacteria		Yeast
	Rhodopseudo-monas spheroides	Rhodospirillum rubrum	Micrococcus lysodeikticus (anaerobic)	E. coli (anaerobic)	
n-C_{15}	2	0·3	112	10	50
n-C_{16}	7	0·7	95	37	60
Pristane	22	3	55	—	—
ΔC_{17}	—	—	—	—	—
n-C_{17}	100	100	100	100	100
Branched C_{18}	—	—	—	—	—
Phytane	3	—	21	—	—
n-C_{18}	44	10	58	700	500
n-C_{19}	43	13	147	210	450
n-C_{20}	8·8	9	53	200	1000
Alkanes of higher mol. wt. (as fraction of total hydrocarbons)	5%	15%	50%	60% of total hydro-carbons	60% of total hydro-carbons
Major component	n-C_{17}	n-C_{17}	n-C_{19}	n-C_{18}	n-C_{20}

Peak heights are relative to n-C_{17} peak taken as 100.

to the bottom of the lake, which is poor in oxygen. If the micro-organisms present in the bottom of the lake are to make use of that carbohydrate they must do so in an anaerobic environment. In order to obtain energy from carbohydrates in an anaerobic environment the organisms have to remove the oxygen from the carbohydrate and dispose of it as carbon dioxide. If the oxygen is being eliminated from the carbohydrates as CO_2, hydrocarbons will accumulate on the bottom of the lake. This idea should be substantiated by a much broader comparative biochemical investigation than has been done so far, in this particular conjecture. If it is true, it will have a very broad application in organic geochemistry.

The other part of our comparative analysis of the existing organisms was an examination of the fatty acids, the molecules with the terminal atoms in the form of a carboxyl (acid) group. There are many of these compounds, and for many years it was believed (and perhaps rightly so) that the odd-dominance of the hydrocarbons found in fairly young petroleums was the result of the decarboxylation of the dominant even fatty acids formed by the two-carbon condensation system. The even

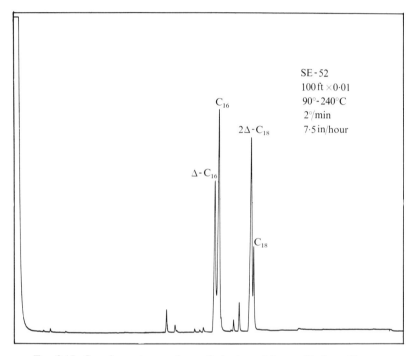

FIG. 3.15. Gas chromatogram for methyl esters of fatty acids from *Nostoc*.

fatty acids, ending in a —CO_2H group, after deposition in the sediments, can lose that CO_2 and form an odd hydrocarbon. This has now been the basic contention of the petroleum chemists for some 20 years, and it is probably to some degree correct, as shown in many of the petroleums that we also have analysed. The question of what the origin of the non-odd hydrocarbons might be is still with us, and we must keep it in mind.

Let us recall the extraction method for the hydrocarbons from the various organisms (algae, bacteria, and yeast); first a benzene–methanol extraction, followed by a heptane extract of the material residue from the evaporation of the benzene–methanol extract. If the fatty acids, which remain unextracted by heptane, are dissolved and esterified with a methylating agent (to make methyl esters from them), they can then be run through the gas chromatographic apparatus almost as easily as the hydrocarbons. The analyses of these fatty acids are shown in Figs. 3.15 and 3.16. Fig. 3.15 shows the esters from *Nostoc*, with a dominance of C_{16}. On the other hand, in the chromatogram for *E. coli* (Fig. 3.16),

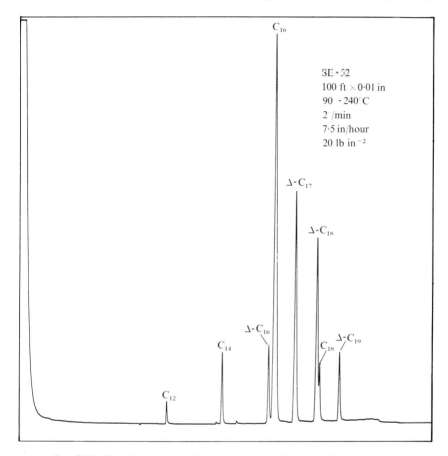

FIG. 3.16. Gas chromatogram for methyl esters of fatty acids from *E. coli*.

while C_{16} is dominant, C_{17}, C_{18}, and C_{19} are also present. The fact that the hydrocarbons in *E. coli* are generally much heavier than the fatty acids in *E. coli* suggests that the hydrocarbons are evolved by a route other than from these acids. At least, the hydrocarbon skeletons are not first made as these acids, which then lose their terminal carboxyl group to become hydrocarbons. This is clear because the hydrocarbon and acid distributions are so different.

Organic geochemistry and ancient sediments

The organic geochemistry of ancient sediments represents an extensive field, in a general sense.[11] There is a very large literature based on the petroleum industry, oriented to two questions. First, the origin of petroleum; i.e. whether it is biological or abiological. For a variety of reasons this is relevant to the search for more oil. If the petroleum is abiological in origin, or if there is any serious component of oil that is abiological, then the deeper we go into the earth, the more oil there should be; if it is solely biological in origin, the reverse would be true. There are schools of thought that consider that there is an abiological component in petroleum and, in fact, that the main body of it has not yet been discovered *in toto*. Only a few of the oils have come from depths at which it might be possible to have an appreciable abiological component. The view that there might perhaps be an abiological origin to some component of petroleum is not so 'unrespectable' as it once was: there are some very respectable chemists who believe it.[12-15] The composition of petroleum is the second question to which the petroleum chemists were oriented, for the sake of knowing what is in the material and what can be done with it chemically. The time is coming when we shall have to use petroleum hydrocarbons solely as chemical raw materials, not merely for fuel, in a much more serious way than has been done in the past. The composition of the raw materials that the chemists will seek to transform into many products (from stockings to drugs) will therefore be important in its intimate detail.

San Joaquin oil (30 million years)

We want to know the detailed molecular composition of the organic constituents in the sedimentary rocks for another reason, and we began this phase of our study with an examination of a relatively young and readily available sample: the San Joaquin oil in central California, which is 30 million years old.[16] We performed gas-chromatographic separations on this material, as shown in Fig. 3.17. Fig. 3.17 (a) shows the total hydrocarbons, after removal of those components of the oil that are not hydrocarbons (i.e. sulphur compounds, nitrogen compounds, etc.) all together. There are no aromatic hydrocarbons in this mixture; these are all alkanes or alkenes. Fig. 3.17 (c) is the gas chromatogram of the normal straight-chain hydrocarbons in the oil, which have been removed from the total mixture by molecular sieving; Fig. 3.17 (b) shows the branched-cyclic hydrocarbons. It is possible to separate the various

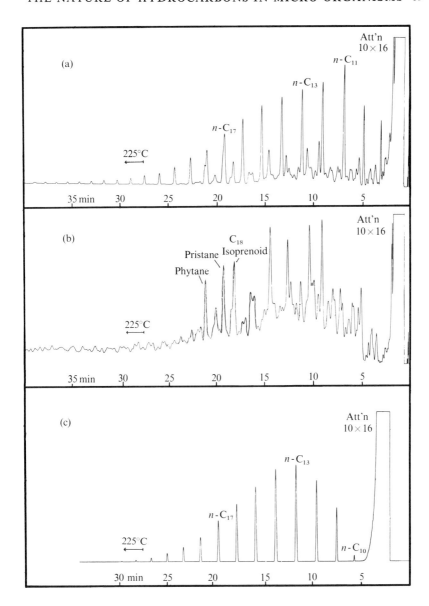

FIG. 3.17. Gas chromatogram of hydrocarbons from San Joaquin oil, central California. (a) Total hydrocarbons; (b) branched-cyclic hydrocarbons; (c) normal straight-chain hydrocarbons.

types of hydrocarbons using synthetic minerals that have holes of the appropriate size and types in them, i.e. molecular sieves. A hydrocarbon without branches or rings can pass into the molecular sieves, leaving behind the branched hydrocarbons, which can be washed out. The mixture of hydrocarbons is shaken up with the molecular sieve material, and time allowed for the long straight-chain hydrocarbons to enter the sieve. Then the branched-cyclic hydrocarbons are washed out and the molecular sieve is dissolved in acid, releasing the straight-chain hydrocarbons, which are then put on the chromatograph. Fig. 3.17 shows no very great odd-over-even dominance. There are, of course, oils that have the odd

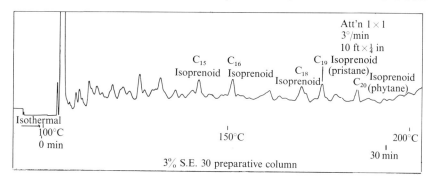

FIG. 3.18. Separation of C_{15}–C_{20} isoprenoids in the branched-cyclic fraction of the San Joaquin oil.

dominance that has been such an important factor in the oil-genesis question, but this is not true to any large extent of the San Joaquin oil. In Fig. 3.17 (b) the isoprenoids are prominent. In addition to the C_{19} and C_{20} peaks, phytane and pristane, which we have discussed, there is also a peak labelled C_{18}. Fig. 3.18 shows the separation of the C_{15}–C_{20} isoprenoids, taken before the two methods of separation and analysis, mass spectrometry and gas–liquid chromatography, had been coupled instrumentally. The mass spectra of the C_{16}–C_{20} isoprenoids are shown in Fig. 3.19. Their behaviour is characteristic and easily recognized and identified, thus confirming the identification of the C_{16}, C_{18}, C_{19}, and C_{20} isoprenoids in the San Joaquin oil. The C_{17} is missing (or at least very rare). The C_{19} and C_{20} (pristane and phytane) are found in nature. We generally do not find the C_{16} and C_{18} so readily in nature; but C_{15} (farnesane) is found in living organisms. It is made by putting three isoprene units together. The next combination of isoprene units would be C_{20}, which would be related to phytol, the alcohol attached to chlorophyll. When we first found phytane in ancient sediments (which was done before we looked at the organisms) we thought it was the result of the abiological reduction and transformation of phytol. It now turns out that we can find phytol itself in the living organism; so it is not so clear that it is only a geological derivative. We do not find the C_{17} hydrocarbon in the San Joaquin oil because in order to go from C_{19} or C_{20} to

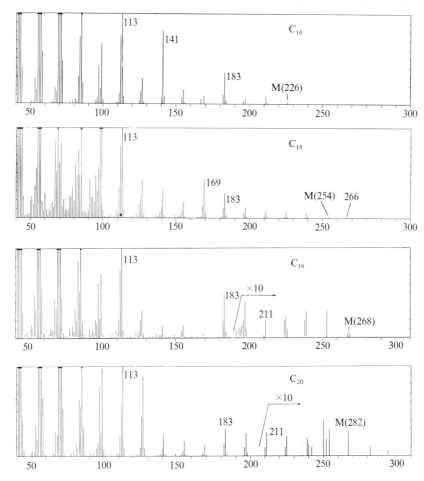

FIG. 3.19. Mass spectra of C_{16}–C_{20} isoprenoids from San Joaquin oil.

C_{18} it is necessary to break only one bond, whereas to go from C_{19} or C_{20} to C_{17} it is necessary to break two carbon–carbon bonds. This may be the reason why C_{17} is so rare: i.e. two carbon breaks are required in order to generate it.

The mass spectra of the San Joaquin oil isoprenoids shown in Fig. 3.19 are clear enough, though somewhat complicated. We shall see still more complex ones later in this section, but all this should be taken as a part of the pattern of recognition that we are trying to establish, so that when we do see something in the ancient rocks we shall have some basis for proper judgement.

Green River Shale (60 million years)

The first rock we studied was the Green River Shale, an oil-rich young sandstone underlying Wyoming and Colorado and a great portion of the plains states.[17] There is a great deal of organic matter in the Green River Shale; about 10 to 20 per cent hydrocarbons. Pl. 1 shows the rock itself, unpolished. The bedding-planes are very clear and are hardly disturbed at all. We have made no attempt to separate the individual layers, but it would obviously be of interest to do this and see whether the chemical composition varies with the time of the deposition, and whether that can be used in any way to determine something about the climate (year by year, for example) during deposition. We do not know what period of geological time those individual bands, each a fraction of a millimetre thick, represent, but I dare say it is more than one season. Pl. 5 shows a polished section of the Green River Shale, approximately 3 in long, and here it is apparent that the individual layers have been disturbed somewhat. There are really three different bands (top, light; middle (disturbed), black; and lower, grey). It looks as though this represents a variation in time over a much longer period, and perhaps this might be the kind of thing in which we could approach the time variation first.

We have taken fragments of the Green River Shale, ground it up, and made our extraction. The gas-chromatographic analysis of the Green River Shale hydrocarbons is shown in Fig. 3.20. The same kind of pattern is visible as was apparent for the San Joaquin oil. We begin to see here the odd dominance; n-C_{17} and n-C_{19} are certainly dominant, as are n-C_{29} and n-C_{31}. There is a bimodal distribution with respect to number of carbon atoms. Thus, for the normal hydrocarbons (Fig. 3.20 (c)) there is a group near C_{17} and another group near C_{30}. A similar distribution appears in the branched-cyclic fraction (Fig. 3.20 (b)). The n-hydrocarbon pattern illustrates the typical odd dominance that the petroleum chemists have been observing for years, which has been interpreted as due to the decarboxylation of the even acids. I am beginning to doubt that theory, but the point is not really central to the present discussion. I also want to call attention to the presence of certain compounds in the branched-cyclic fraction of the Green River Shale, labelled

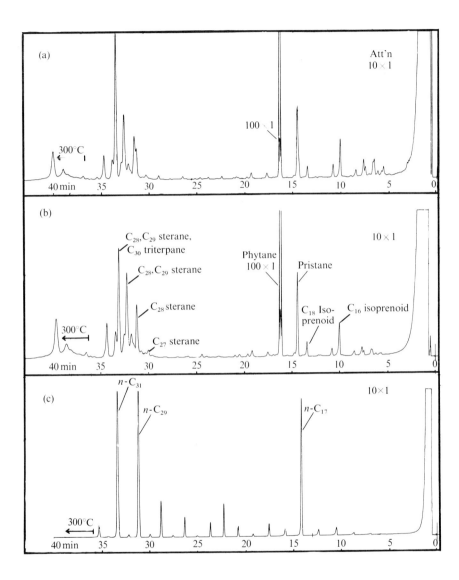

FIG. 3.20. Gas chromatogram for total hydrocarbons from Green River Shale, Colorado (Eocene, approx. 60 million years old). (a) Total hydrocarbons; (b) branched-cyclic fraction; (c) normal fraction.

C_{27} (sterane), C_{28} (sterane), C_{29} (sterane), and C_{30} (carotene). Keep in mind the pattern of the steranes, which are cyclic compounds. Fig. 3.21 shows the capillary chromatogram of the group of compounds in the Green River Shale between the C_{16} isoprenoid and the C_{20} phytane. In this high-resolution chromatogram it is clear that this particular group will be useful in our further studies and that it is not as simple as it at first appeared.

Returning to the steranes in the Green River Shale, I want to relate that particular group to those we have already discussed in the isoprenoids. The steranes are hydrocarbons closely related to the isoprenoids. The sterane general class formula is included in Fig. 3.22 (every corner and end represents a carbon atom). Fig. 3.22 shows the mass spectra of these three sterane hydrocarbons C_{27}, C_{28}, C_{29} compared with a sample of authentic sitostane. The C_{27}, C_{28}, and C_{29} mass spectra are all dominated by the 217 peak, which is also found in the authentic sample of sitostane. The dominance of the 217 peak is a fragmentation that tells us, since we see the same 217 mass peaks in all three of these steranes, that the homology, that is the twenty-eighth and twenty-ninth carbon atom, is in the part of the molecule that came off and not in the part retained in the ion of mass 217.

ionization at quaternary carbon

allylic carbonium ion
$M = 217$ $(C_{16}H_{25})$

Another piece of information comes from the 149 mass peak fragment (which is similar to the information that comes from the 217 peak), showing that the hydrocarbon homology is in the side chain:

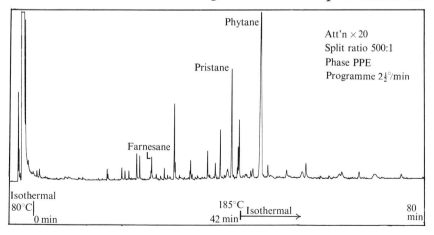

$$M = 149 \ (C_{11}H_7)$$

If that is the case, we ought to be able to discover a fragmentation that has resulted from the breaking-off of a common piece such as the

FIG. 3.21. Capillary chromatogram of the lower molecular weight total hydrocarbon alkane fraction from the Green River Shale (Eocene).

A ring, leaving the charge with the homologous fragments. There is a break corresponding to a loss of eight carbon atoms from each of the three molecules; and that particular break moves up one carbon atom in the charged fragment as you go along.

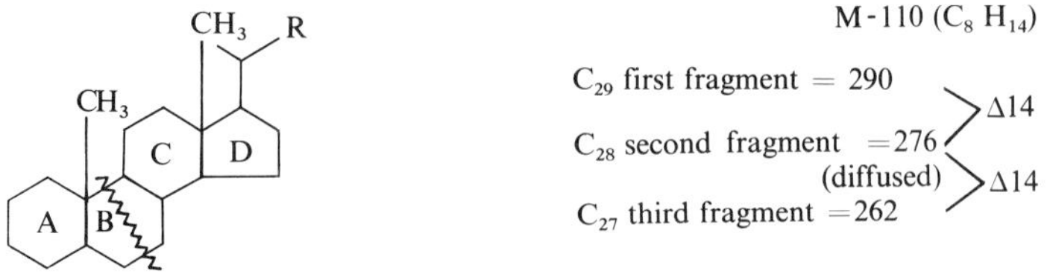

CH3
R
CH3
C
D
A
B
M-110 (C8 H14)
C29 first fragment = 290
C28 second fragment = 276
(diffused)
C27 third fragment = 262
Δ14
Δ14

The evidence is fairly clear that sitostane is the C_{29} hydrocarbon with which we are dealing among the steranes from the Green River Shale.[18] The meaning of the particular homology remains to be determined. We do, however, have a series of cyclic compounds that are very complex, that surely did not evolve to dominance by some abiogenic accident, and because of their dominant position are almost surely the result of biological activity. This group of ring compounds is closely related to the open-chain isoprenoids through squalene (C_{30}) in the following way:

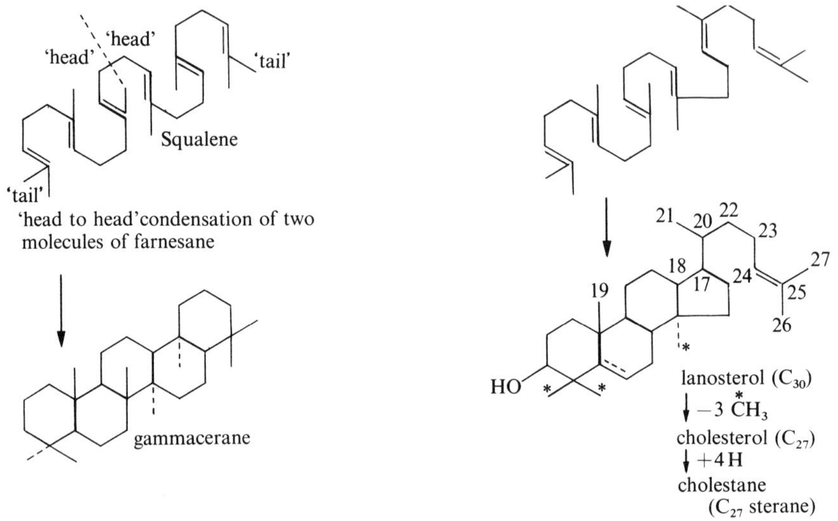

'head'
'head'
'tail'
Squalene
'tail'
'head to head'condensation of two molecules of farnesane
gammacerane
21
20
22
23
18
24
27
19
17
25
26
HO
lanosterol (C30)
−3 CH3
cholesterol (C27)
+4H
cholestane
(C27 sterane)

We have already seen in a general way how the isoprenoids are constructed. We can go on to the 15-carbon unit made of three of the isoprene units. Two of the 15-carbon units (farnesane) put together, head to head, give squalene, a C_{30} compound, found in shark's liver among other places. It has been shown that squalene is a precursor to the ring-closed compounds and, in fact, one particularly symmetrical pentacyclic six-ring compound, gammacerane, was isolated from the

Green River Shale. The other route for ring closure would be via three six-ring and one five-ring closure, followed by the loss of three methyl groups. This gives the other structure, a C_{27} sterane, cholestane, also present in the Green River Shale. This is the route by which such compounds are generated in living organisms. Therefore, these ring compounds, which look rather different from the open-chain isoprenoids,

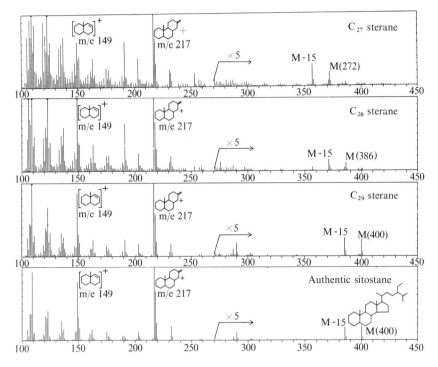

FIG. 3.22. Mass spectra for three sterane hydrocarbons, C_{27}, C_{28}, C_{29}, and, for comparison, authentic sitostane. The general formula for the steranes is shown.

are, in fact, very closely related and if we find both of them in the same sample we are quite convinced that the particular sample is of biological origin.[19-22]

We shall now continue our discussion of some of the more ancient rocks. We shall go back from 60 million years (the Green River Shale) to roughly 1000 million years and leave out two intermediate ages. These introduce some new homologies, new hydrocarbon series, such as those with a single methyl group at the end and those with a methyl group two carbon atoms in from the end (the iso and anteiso series), but we shall see those in the 1000-million-year-old rocks as well.

REFERENCES

1. HAN, J., Ph.D. thesis, Department of Chemistry, University of California, Berkeley.
2. —— McCARTHY, E. D., VAN HOEVEN, W., CALVIN, M., and BRADLEY, W. H., Organic geochemical studies. II. The distribution of aliphatic hydrocarbons in algae, bacteria and in a recent lake sediment: a preliminary report. *Proc. natn. Acad. Sci. U.S.A.* **59**, 29 (1968).
3. CORNFORTH, J. W. and RYBACK, G., The stereospecificity of enzymic reactions. *Rep. Prog. Chem.* **42**, 428 (1967).
4. POPJAK, G. and CORNFORTH, J. W., The biosynthesis of cholesterol. *Adv. Enzymol.* **22**, 281 (1960).
5. —— Some aspects of lipid biochemistry. *Proc. R. Soc.* **B156**, 376 (1962).
6. BLOCH, K., The biological synthesis of cholesterol. *Science* **150**, 19 (1965).
7. LYNEN, F., Biosynthesis of saturated fatty acids. *Fed. Proc.* **20**, 941 (1961).
8. HOERING, T. C., Criteria for suitable rocks in Precambrian organic geochemistry. *Carnegie Institution Yearbook*, Vol. 65, p. 365 (1966).
9. HAN, J., McCARTHY, E. D., CALVIN, M., and BENN, M. H. The hydrocarbon constituents of the blue-green algae, *Nostoc muscorum, Anacystis nidulans, Phormidium loridum* and *Chlorogloea fritschii. J. chem. Soc.* In press.
10. McCARTHY, E. D., HAN, J., and CALVIN, M., Hydrogen atom transfer in the mass spectrometric fragmentation patterns of saturated aliphatic hydrocarbons. *Analyt. Chem.* **40**, 1475 (1968).
11. For a general review on the subject, see HOERING, T. C., The organic geochemistry of Precambrian rocks. *Researches in organic geochemistry* (editor P. H. Abelson), Vol. 2, p. 87. New York (1967).
12. ROBINSON, SIR ROBERT, The origins of petroleum. *Nature, Lond.* **212**, 1291 (1966).
13. —— Origins of oil: A correction and further comment. Ibid. **214**, 263 (1967).
14. SYLVESTER-BRADLEY, P. C. and KING, R. J., *Evidence for abiogenic hydrocarbons.* Ibid. **198**, 728 (1963).
15. HENDERSON, W., EGLINTON, G., SIMMONDS, P., and LOVELOCK, J. E., Thermal alteration as a contributory process to the origins of petroleum. Ibid. **219**, 1012 (1968).
16. JOHNS, R. B., BELSKY, T., McCARTHY, E. D., BURLINGAME, A. L., HAUG, PAT, SCHNOES, H. K., RICHTER, W., and CALVIN, M. Organic geochemistry of ancient sediments. II. *Geochim. Cosmochim. Acta* **30**, 1191 (1966).
17. CALVIN, M., Chemical evolution (the Bakerian lecture). *Proc. R. Soc.* **A288**, 441 (1965).
18. BURLINGAME, A. L., HAUG, PAT, BELSKY, T., and CALVIN, M., Occurrence of biogenic steranes and penta-cyclic triterpanes in an Eocene shale (22 million years) and in an Early Precambrian shale (2·7 billion years): a preliminary report. *Proc. natn. Acad. Sci. U.S.A.* **54**, 1406 (1965).
19. HAUG, PAT, SCHNOES, H. K., and BURLINGAME, A. L., Isoprenoid and dicarboxylic acids isolated from Colorado Green River shale (Eocene). *Science* **158**, 772 (1967).
20. BURLINGAME, A. L. and SIMONEIT, B. R., Isoprenoid fatty acids isolated from the kerogen matrix of the Green River formation (Eocene). Ibid. **160**, 531 (1968).
21. EGLINTON, G., DOUGLAS, A. G., MAXWELL, JAMES R., RAMSAY, J. N., and

STÄLLBERG-STENHAGEN, S., Occurrence of isoprenoid fatty acids in the Green River shale. Ibid. **153,** 1133 (1966).

22. MACLEAN, IAIN, EGLINTON, GEOFFREY, DOURAGHI-ZADEH, K., ACKMAN, R. B., and HOOPER, S. N., Correlation of stereoisomerism in present-day and geologically ancient isoprenoid fatty acids. *Nature, Lond.* **218,** 1019 (1968).

4

MOLECULAR PALAEONTOLOGY

WE have completed our examination of the starting materials, i.e. the bacteria and algae that might conceivably have produced the starting materials that are today fossilized in the rocks. We have looked at some of the early steps in the formation of the sediments, taking as an example the Mud Lake in Florida. We have also examined two rather young hydrocarbon sources: a California (San Joaquin) oil, 30 million years old, and an oil shale, the Green River Shale, 60 million years old. We have found certain characteristic patterns of the hydrocarbon distribution in these various materials that seem, at least, to give evidence of biological origin.

We could continue this progression, examining successively older rocks each time going through the same sequence of analyses (extraction, separation of saturated hydrocarbons from benzenoid and olefinic (or unsaturated) hydrocarbons, separation of saturated hydrocarbons into their various component parts by gas–liquid chromatography and mass spectrometry). I do not think, however, that this would gain us much, for two reasons. One is that it would be boring to go through the samples one after another, coming out with more or less the same patterns, only slightly different each time. And there is really no need to take up space in going through this laborious progression, even though there are some systematics to be derived from this kind of study that will ultimately be useful in helping us to attain our primary objective.

I should therefore like to pass directly to the ancient rocks, those beyond the 1000-million-year mark, and summarize at the end of that study what little systematics can so far be derived from such an examination. The end-result will serve just as well.

Molecular palaeontology of ancient rocks

The ancient rocks we shall now discuss are those described in the first chapter in which 'microfossils' are present: the Nonesuch Shale from

Michigan, 1000 million years old, the Gunflint Shale from Michigan, 1900 million years old, the Soudan Shale, from Michigan and the Canadian Shield, 2700 million years old, and the Fig Tree Chert from southeast Africa, dated at 3100 million years.

Since there are a number of deductions to be made about the origin of a molecule from a knowledge of its intimate architecture and milieu (molecular and geological), we shall occasionally diverge from the direct purpose to examine some detailed architecture that is important for the main question. These diversions will take place only when the experimental data warrant it, and when the information that such a detailed examination will provide is really pertinent to the main issue.

Nonesuch Shale (1000 million years)

The Nonesuch Shale is found in the old White Pine iron mine in northern Michigan.[1] There was, until a few decades ago, an extensive

Fig. 4.1. Nonesuch Shale (Pre-Cambrian) with calcite vein containing petroleum. From White Pine mine, Michigan. Age, 1×10^9 years.

iron-mining industry in deep mines in northern Michigan. This has since been superseded by the open-pit mining of Minnesota and Labrador, and the old mines are gradually being closed. The White Pine mine has not been used for some time. I am afraid we can no longer get samples from it, at least not from the deep levels, which have collapsed.

A photograph of a hand-specimen of the Nonesuch Shale is shown in Fig. 4.1. The shale, the marker bed itself, is the bottom; above it is an

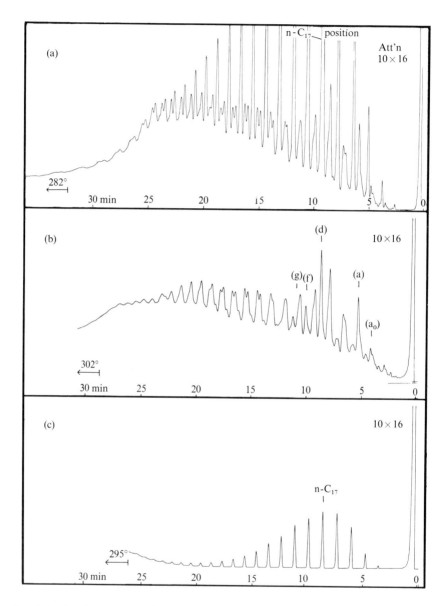

FIG. 4.2. Gas chromatogram for alkane fractions from oil in Nonesuch Shale. (a) Total alkanes; (b) branched-cyclic hydrocarbons; (c) normal hydrocarbons.

oil seep; and above that is some calcite that contains oil enclosed in the calcium carbonate crystal. We have examined each of these crude fractions separately, with the object of seeing if there was any difference in the hydrocarbon content of the three different materials. The question was to determine whether or not there was any separation of the basic hydrocarbons in the course of this stratification. One component is the seep oil; another is the oil that is encased in the limestone; and a third is the oil that is entrapped in the shale itself and can be extracted only by grinding the rock. Fig. 4.2 shows the gas chromatogram of the oil fractions (alkane fractions).[2,3] In the normal fraction (Fig. 4.2 (a)) there is a peak at about C_{17}, and no odd-even dominance or odd-even alternation as in the Green River Shale. There is a homology in the higher numbers of the branched-cyclic fraction, which we believe to be of the iso and anteiso alkanes, some of whose structures are as yet undetermined. Fig. 4.3 shows the gas-chromatographic separation of the alkane fractions of the calcite vein of the Nonesuch Shale, which is quite similar to the distribution shown in Fig. 4.2. Fig. 4.4 illustrates the gas chromatogram of the fractions from the shale itself (the marker bed); this shows a normal distribution peaking a little lower than C_{17}. Fig. 4.5 (a) shows the total alkane group on a high-resolution capillary chromatogram: each of the peaks is a different compound. The combined group of fractions lying between a_0 and g is shown in Fig. 4.5 (b). Authentic samples of farnesane (the C_{15} isoprenoid), pristane (the C_{19} isoprenoid), and phytane (the C_{20} isoprenoid) have been added. It is clear that all three of these materials are present in the Nonesuch Shale. There are also many unidentified compounds. I presume that when the intimate molecular architecture of some of these unknown compounds is finally determined it will be of importance in determining the origin of the Nonesuch Shale.

Gunflint Shale (1900 million years)

The Gunflint Shale is a clay-containing shale found in Michigan.[4] Fig. 4.6 shows a gas chromatogram of the heptane-soluble and total alkane fractions from the Gunflint Argillite. Fig. 4.6 (a) shows the heptane-solubles, i.e. a fraction from which nothing has been removed and which contains the aromatic hydrocarbons as well as the unsaturated hydrocarbons. Fig. 4.6 (b) is the gas chromatogram of the total alkanes, showing the dominance of the normal hydrocarbons, although it is clear that there are also some sterane-types, or at least some fairly heavy materials, present.[5] Fig. 4.7 shows the gas chromatogram of the Gunflint Chert, a siliceous rock of the Gunflint formation, which gives the same

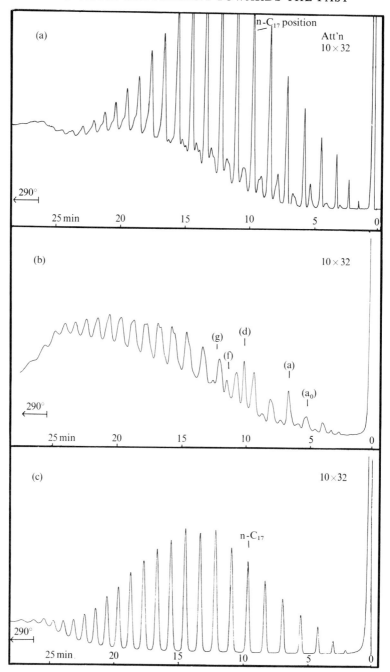

FIG. 4.3. Gas chromatogram for alkane fractions from calcite vein in Nonesuch Shale. (a) Total alkanes; (b) branched–cyclic hydrocarbons; (c) normal hydrocarbons.

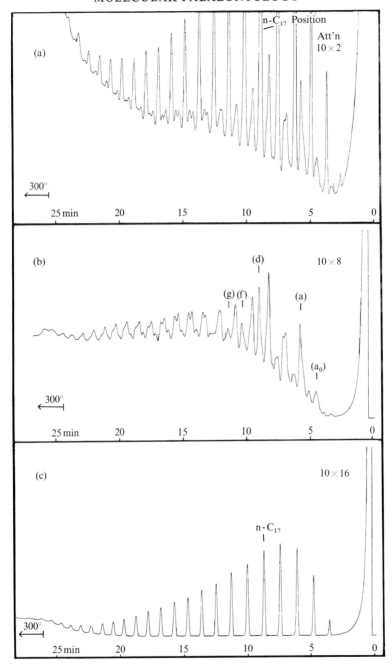

FIG. 4.4. Gas chromatogram of alkane fractions from Nonesuch Shale marker bed. (a) Total alkanes; (b) branched–cyclic hydrocarbons; (c) normal hydrocarbons.

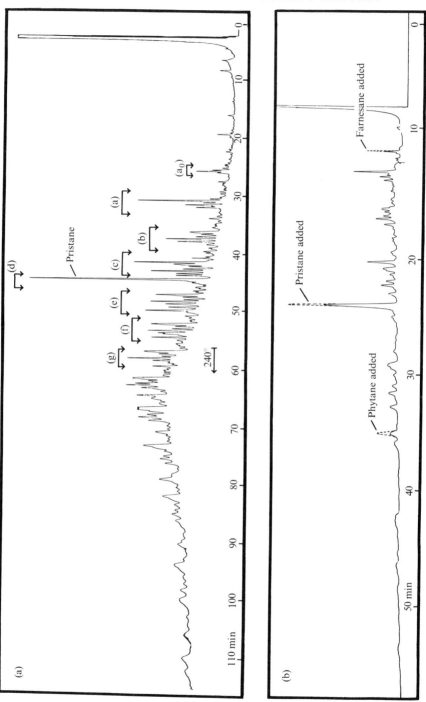

FIG. 4.5. High-resolution capillary chromatograms of alkanes from oil in Nonesuch Shale. (a) Total fraction; (b) combined group of fractions from a_0 to g.

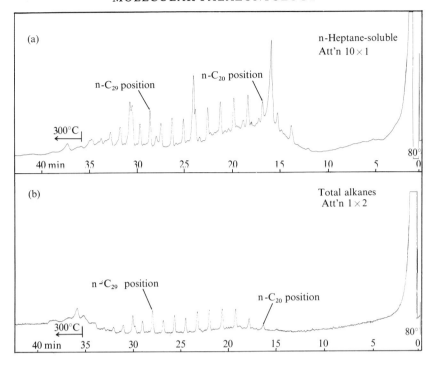

FIG. 4.6. Gas chromatogram of heptane-soluble and total alkane fractions from Gunflint Argillite. (a) Heptane-soluble fraction; (b) total alkanes.

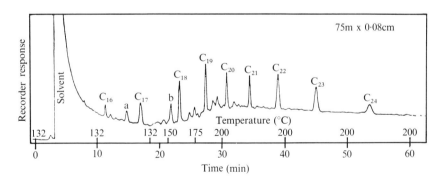

FIG. 4.7. Gas chromatogram of alkanes from Gunflint Chert. *a* is probably pristane and *b* phytane.[6]

kind of distribution, with the normal hydrocarbons fairly well distributed.[6] There are two extra peaks, one just ahead of the n-C_{18} and one just ahead of the n-C_{17}. Judging from what we have seen of this kind of chromatography in other instances, the one labelled a is almost surely pristane, the C_{19} isoprenoid, and the peak labelled b (just ahead of the n-C_{18}) is undoubtedly phytane, the C_{20} isoprenoid. Here in the Gunflint Chert, 1900 million years old, we see again a similar mixture: the straight-chain hydrocarbons with two or three of the important isoprenoids among them.[7]

Soudan Shale (2700 million years)

A sample of the Soudan Shale is shown in Fig. 4.8.[8] On the left-hand side is a graphitic inclusion in the rock, and on the right is the clay-like

FIG. 4.8. The Soudan iron formation: left, graphitic lens; right, carbon-rich argillite.
Pre-Cambrian; age approx. 2.5×10^9 years.

shale of the iron-formation of the Soudan rock. Fig. 4.9 shows the hard shale of the Soudan, a very dense substance, sliced. Fig. 4.10 is the gas chromatogram of the alkane fractions of the Soudan Shale showing the dominance of the isoprenoids. The isoprenoid range is somewhat extended and here we see something new coming in, a C_{21} isoprenoid. In the branched–cyclic fraction (shown in Fig. 4.10 (b)) is the group of compounds called the steranes, which were also found in the Green River

Shale and are the tetracyclic alkanes made from isoprenoids.[9,10] The gas chromatograms shown in Fig. 4.10 were done on Apiezon L, which is a particular kind of coating in the chromatographic tube; a higher-resolution chromatogram, done on PPE (polyphenylether), is shown in Fig. 4.11. The C_{21} isoprenoid identified in this Soudan Shale chromatogram is 2, 6, 10, 14-heptadecane. This was done by co-chromatography

FIG. 4.9. Soudan Shale. The phial contains the heptane fraction from the rock.

and mass spectrometry. The mass spectrum alone does not really distinguish between three alternative possibilities for this 21-carbon atom isoprenoid. The reason why it is important for us to know the detailed structure of the C_{21} isoprenoid will become apparent when we look at the possible alternative structures that might be assigned to it:

(a)

C_{21} isoprenoid.

Note that with 21 carbon atoms we have now passed beyond the point at which we can derive the compound from phytol, which is the alcohol fragment attached to chlorophyll. Phytol is a normal C_{20} isoprenoid

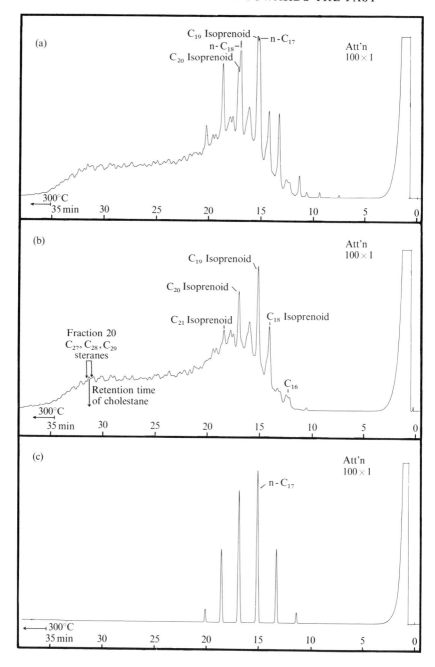

FIG. 4.10. Gas chromatograms of alkanes from Soudan Shale (surface sample). (a) Total alkanes; (b) branched–cyclic alkanes; (c) normal alkanes.

made up of four C_5 fragments hooked together head to tail in the normal fashion, giving 20 carbon atoms in a normal isoprenoid configuration. What then is the C_{21} hydrocarbon? It is clear from the mass spectrum that it is an isoprenoid, but the mass spectrum cannot tell us precisely where the extra carbon atom is. It might be on the head end (the 1, 2, 3 carbon end above); or the branch might not be at the fourteenth carbon as above but at the fifteenth carbon atom, giving a two-carbon tail end and a four-carbon fragment (11, 12, 13, and 14) just ahead. There are thus at least three different possible structures. If there is a four-carbon piece in the middle, i.e. if the compound were 2, 6, 10, 15-heptadecane, then with the four carbons in the middle it could be derived from the very common squalene, which has four carbons in the middle and is made by hooking together C_{15} compounds, head to head, making a C_{30} hydrocarbon with a four-carbon group between the two branch points in the centre. It is conceivable that such a C_{21} isoprenoid could be a fragment of squalane.

(b)

C_{30} squalane.

On the other hand, if the C_{21} isoprenoid contains a three-carbon end, with only a three-carbon interval between the two branch points, it cannot be derived from squalane, nor can it come from phytol (phytane). A lot of effort was spent in getting information that might help to determine the origin of the C_{21} isoprenoid. We synthesized two of the possible compounds, i.e. the 2, 6, 10, 14-heptadecane and the 2, 6, 10, 15-heptadecane; we were able to show that the mass spectra of the two compounds were very similar, i.e. insufficiently different to be used as the sole means of identification.[11,12] However, the two different synthetic 'models' could be separated by capillary gas chromatography, and Fig. 4.12 shows the separation of the 2, 6, 10, 14, and the 2, 6, 10, 15-heptadecanes. The co-chromatography thus provides a significant identification of these two structures and the distinction between them.

As the C_{21} isoprenoid in the Soudan Shale has been identified as 2, 6, 10, 14-heptadecane, it obviously cannot originate with phytane, which has only 20 carbon atoms; it cannot derive from squalene or squalane (ordinary fish oil); it must have some other precursor. It turns out that there are no normal isoprenoids known in nature in the C_{25} range. One might think that if there were a C_{20} isoprenoid, that is four isoprenoid units hooked together in the normal fashion, there might very well be one with five, a C_{25}. However, we have to go up to C_{45}, solanosane (solanosol), before we begin to see again the normal isoprenoids in which these five-carbon units are hooked together in a uniform head to tail fashion, with a normal isopropyl head end.[13,14]

Solanosane is a likely source for a C_{21} hydrocarbon of the type found in the Soudan Shale. The solanosane would break at one of the tertiary carbon atoms near its centre, where the normal break is readily achieved, to give the compound with the three-carbon tail on it; and this is what we actually find. Solanosane is a relatively rare material, and has been described as present in only a few organisms.

The C_{21} isoprenoid in the Soudan Shale might also come from lycopene, a very common material that is the precursor of vitamin A.[15] Lycopene has a 40-carbon skeleton, which could give rise to the C_{21} by cracking at the C_{17} junction.

That, however, is an unusual place for such a compound to break; it would have to break between two secondary carbon atoms, not at one of the tertiary branch points, which is the more usual breaking-point. My inclination is to believe that the C_{21} isoprenoid in the Soudan Shale does not come from lycopene, although this is a plentiful compound in nature, but rather from something more like solanosane. The discovery of bactoprenol in *Lactobacilli* is suggestive.[16] We must search for the possible precursors for the C_{21} with some care in the future.

Thus the identification of the C_{21} isoprenoid from the Soudan Shale as 2, 6, 10, 14-heptadecane raises some question of its origin; it *cannot* have been derived either from phytane (the common isoprenoid associated with photosynthetic organisms) or from squalane (the common hydrocarbon in marine organisms).[17]

Fig. 4.13 shows the gas chromatogram of a Soudan Shale branched–cyclic fraction on Apiezon L (a different kind of absorbent in the gas-chromatographic column). In Chapter 3 we considered a rather special hydrocarbon, the so-called branched C_{18}, which was peculiarly structured and found in *Nostoc*.[18] The branched C_{18} compound was assigned the structure of an equimolar mixture of 7-methyl and 8-methyl heptadecanes. This is a very peculiar branched C_{18} compound, a C_{17} straight chain with branches on the seventh and eighth carbon atoms respectively—and present in equal amounts. The blue-green alga *Nostoc* is one of the types of microfossil that is believed to be present in some of the

(c) C$_{45}$ solanosane.

(d) C$_{40}$ lycopane.

ancient rocks, though not in the Soudan Shale. Nevertheless, as soon as the question of the structure of this branched C_{18} compound in *Nostoc* was satisfied up to a point, the next step was to find out whether there was any evidence that this peculiar structure was present in the Pre-Cambrian rocks themselves. The hydrocarbon (b-C_{18}) from the *Nostoc*

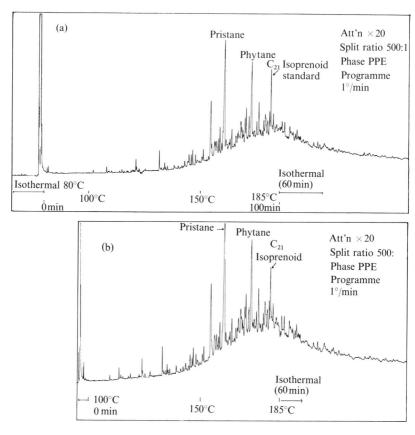

FIG. 4.11. High-resolution gas chromatograms of alkanes from Soudan Shale. (a) Branched–cyclic fraction and C_{21} isoprenoid standard; (b) branched–cyclic fraction detail.

was purified, co-chromatographed with various fractions of the Soudan Shale and on a variety of fractionating materials, that is, a variety of bases on the chromatographic apparatus—Apiezon L, SE–30 (a silicone grease), and PPE (polyphenylether). It turned out that a particular peak in Fig. 4.13 seems to co-chromatograph with the b-C_{18} out of the *Nostoc*. This is a very curious structure. If this peculiar mixture is in fact shown to be present in the Soudan Shale, then we shall have another kind of 'molecular marker' that is more special, more sharply defined, than anything found so far. This will become of much more importance in even more ancient rocks.

Thus, the Soudan Shale may contain in it this peculiar branched C_{18} hydrocarbon that is present in the blue-green alga *Nostoc*.

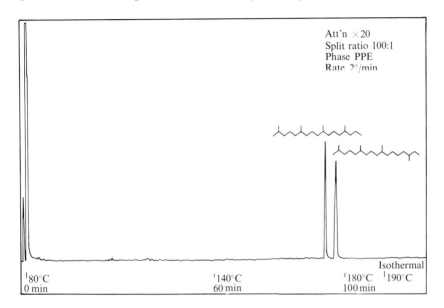

Att'n ×20
Split ratio 100:1
Phase PPE
Rate 2°/min

| 80°C | 140°C | 180°C | Isothermal |
| 0 min | 60 min | 100 min | 190°C |

FIG. 4.12. Capillary gas chromatogram showing order of elution of C_{21} isoprenoid isomers (2, 6, 10, 14-heptadecane and 2, 6, 10, 15-heptadecane).

Fig Tree Shale (3100 million years)

The Fig Tree Shale (Fig. 4.14) is black and has a great deal of carbon in it, mostly insoluble.[19] However, we and others were able to extract a significant amount of soluble carbon from specimens.[20] There is no great dominance of any particular hydrocarbon, and Fig. 4.15 shows a fairly uniform distribution between the n-C_{14} to n-C_{25} hydrocarbons. There is also evidence for the probable presence of pristane at point *a* in Fig. 4.15 and probably also phytane, just ahead of the n-C_{18}. In this case there is no dominance of odd over even chains.

Onverwacht Series (3700 million years)

The most ancient rock so far alleged to contain microfossils has just been described.[21] Recently I had occasion to see the first chromatogram of the hydrocarbons obtained from this sample of rock, the Onverwacht (Fig. 4.16).[22] Here we see a rather spectacular departure from any of

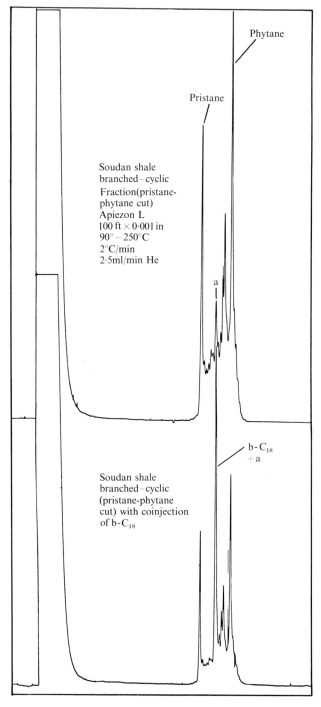

FIG. 4.13. Gas chromatogram of total alkanes from Soudan Shale on Apiezon L.

ERRATA

The sample of Green River shale (Wyoming) shown in Plate 1 has been polished. The sample of the same shale shown in Plate 5 is unpolished.

PLATE 1

Green River Shale (Wyoming) (60×10^6 years)

PLATE 2

Nonesuch Shale (Michigan) (1×10^9 years)

FIG. 4.14. Hand-specimen of Fig Tree Shale. Pre-Cambrian; age approx. $3 \cdot 1 \times 10^9$ years.

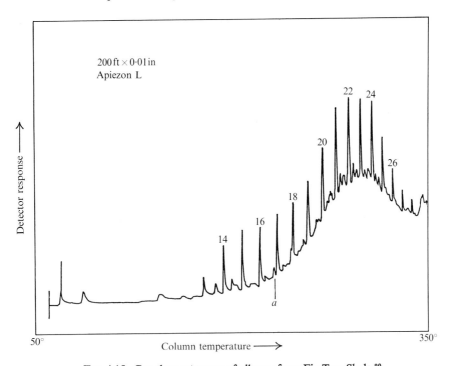

FIG. 4.15. Gas chromatogram of alkanes from Fig Tree Shale.[20]

the previous rock hydrocarbon samples we have ever seen. It appears that there is a very large number of relatively low molecular weight hydrocarbons, ranging from n-C_{16} to n-C_{25}, so various as to give what appears to be a continuum of depositions. Superimposed on this continuum are a number of rather specific peaks which we have, up until now, associated with biogenic origin. This is precisely the kind of spectrum one might

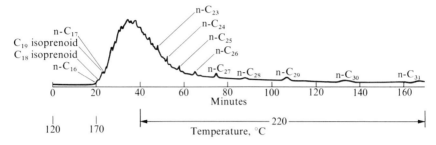

Fig. 4.16. Onverwacht Series sedimentary rock saturated hydrocarbon extract, temperature programmed on a 15 ft × 0·020-in i.d. SCOT column coated with Apiezon L. (W. Macleod, UCSD.)

expect from a mixture of abiogenic materials (the continuum) and biogenic materials (the superimposed peaks). If this is indeed the case (and further examination of the Onverwacht hydrocarbons by the most sensitive of gas-chromatographic and mass-spectrometric methods must be accomplished before this can be decided) we may indeed have found a period in time, 3700 million years ago, during which the transition between the abiogenically developed organic substrate was being converted by the newly developed autocatalytic replicating chemical systems which give rise to life.

Carbon isotope ratios as criteria for biological origin

Evidence of a quite different kind has been introduced as a criterion of biological origin of the carbon in the ancient rocks. Many years ago, when we first began our work on photosynthesis, we carried out an experiment in which we put radioactive carbon dioxide in a growth chamber that had a radioactivity analyser to measure $^{14}CO_2$ and an infrared analyser on the gas stream to measure the total CO_2. We soon found that there was a discrepancy between the two measurements. There was evidence that the green plant was selectively absorbing the lighter carbon dioxide rather than the heavier one; that is, the infra-red resonance absorption was set on the $^{12}CO_2$ and did not 'see' the ^{14}C, whereas the ionization chamber did not 'see' the ^{12}C and analysed only the ^{14}C. In effect, the plant was an 'isotope separator', concentrating the heavy isotope in the remaining gas phase. This is now known to be a generalized phenomenon—the green plants select the light isotope.[23,24] In fact, there is now a whole history of carbon-12/carbon-13 isotope analysis in an attempt to use the natural variation in the isotope as some clue to the history of that particular carbon, in whatever form it might be. The natural abundance of the isotope can be expressed as follows. The ratio of the ^{13}C to ^{12}C is measured in a sample of CO_2, and the instrument is quickly shifted over to some standard carbonate. By running the determinations alternately, the difference between sample and standard can be measured accurately, to parts per thousand. The deficiency of the heavy isotope is simply defined in the following way:

$$\delta^{13}C = \frac{(^{13}C/^{12}C)_x - (^{13}C/^{12}C)_{\text{stand}}}{(^{13}C/^{12}C)_{\text{stand}}} \times 1000,$$

where the subscripts x and 'stand' refer to the sample and standard respectively. A particular marine carbonate is the standard of isotope measurement. Most samples are more deficient in ^{13}C than the standard, and in consequence the scale is very nearly all on the negative side. Most natural processes are systems that discriminate against the heavy isotope, and they therefore result in a greater deficiency of carbon-13. This type of measurement and effect is shown in Fig. 4.17.[25] Even the distribution of carbon dioxide between the atmosphere and the ocean bicarbonate involves a fractionation. Marine algae and marine invertebrates increase the heavy carbon isotope deficiency by another 10 to 15 parts per thousand; coal is down another 20 parts per thousand in the isotopic content, and land plants are about the same. Petroleum, both marine and freshwater, is clearly deficient in heavy carbon. We then looked at the isotopic composition of these ancient rock hydrocarbons; this was done by Hoering. Tables 4.1 and 4.2 show some of the $^{13}C/^{12}C$ ratios in the organic matter of ancient rocks.[26] This deficiency is in about the proper ratio if these materials are indeed of biological origin. We do not know, of course, what would happen if we generated these hydrocarbons, particularly the soluble hydrocarbons, from methane by some abiological mechanism as yet unknown to us. That experiment has not yet been done properly.

There are a number of discrepancies between the soluble and insoluble carbon in these ancient sediments. Some people feel that the soluble carbon and insoluble carbon in the Soudan Shale are different in origin. Hoering, who made these measurements, holds this view. He felt that the

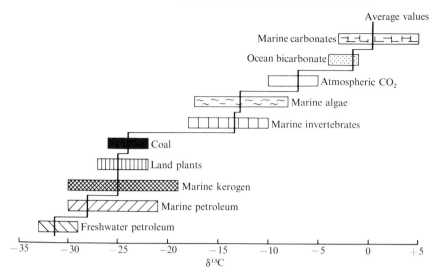

FIG. 4.17. Carbon-13 heavy isotope deficiencies (^{13}C) in various types of organic matter. (After Craig 1953; Wickman 1952; Silverman and Epstein 1958; Eckelman *et al.* 1962; Silverman 1962, 1963; Degens 1965; and others.)[25]

TABLE 4.1

^{13}C/^{12}C *ratios in the soluble and insoluble organic matter of four unmetamorphosed Pre-Cambrian rocks*

Rock	Location	^{13}C soluble organic matter	^{13}C insoluble organic matter
Shale, Nonesuch Formation	Michigan, U.S.A.	−28·14	−28·15
Shale, McMinn Formation, Roper River Series	Northern Territory, Australia	−30·59	−30·71
Shale, Muhos Formation, Jotnian Series	Finland	−27·51	−28·71
Shale, Jeerinah Formation, Fortesque Series	Western Australia	−24·11	−36·50

The units used are defined on p. 89. The standard is National Bureau of Standards isotope reference material No. 20, limestone from Solenhofen, Bavaria.

discrepancies might indicate that they come from different sources. I do not believe this is so. It is easy enough to see that if we started with any normal hydrocarbon and subjected that isotopically normal hydrocarbon to any kind of process that would break carbon–hydrogen bonds giving the dehydrogenated insoluble kerogen, the carbon–hydrogen bond most readily broken would be a ^{12}C—H rather than ^{13}C—H, because of the isotope effect. This would mean that the carbon-12 would accumulate in dehydrogenated material, and therefore, that this material would be carbon-13 deficient. Thus, for an observational example, it makes no difference in which direction we change the normal isotopic content of the C_{28} fraction in a petroleum; i.e. breaking either a carbon–carbon or

TABLE 4.2

$^{13}C/^{12}C$ *ratios in the soluble and insoluble organic matter of some metamorphosed Pre-Cambrian rocks*

Rock	Location	^{13}C soluble organic matter	^{13}C insoluble organic matter
Shale, Soudan Formation	Michigan, U.S.A.	$-25{\cdot}00$	$-34{\cdot}81$
Carbon leader, Witwatersrand system	South Africa	$-27{\cdot}36$	$-35{\cdot}22$
Shale, Fig Tree Series, Swaziland System	South Africa	$-27{\cdot}55$	$-26{\cdot}94$
Shale, Ventersdorp System	South Africa	$-25{\cdot}78$	$-36{\cdot}86$
Carbonaceous limestone, Transvaal System	South Africa	$-25{\cdot}01$	$-38{\cdot}21$

a carbon–hydrogen bond. Breaking a C—C bond results in lighter hydrocarbons; breaking a C—H bond results in insoluble hydrocarbons, and both of these resulting materials should be more deficient in carbon-13 than the starting material, as indeed they are. Thus, the isotopic composition of the small-molecule fraction in the petroleum (methane, ethane, propane, butane, pentane) is deficient in ^{13}C, but up at about C_{28} the isotopic composition approaches normal, and for the materials of high molecular weight (the less soluble residues) the ^{13}C level drops again. It seems to me that it is not a strong or compelling argument to say that because there is a difference between the soluble and insoluble carbon isotope deficiencies the two materials must have different origins.

Organic geochemistry—summary

I should like, in summary, to review a variety of evidence that living organisms might have existed as early as 3100 million years ago. The evidence is really this: a particular variety of rather specific types of hydrocarbons has been found ranging all the way back into Archaean times as far back as the Fig Tree Shale.[27] This particular group is shown in Fig. 4.18, ranging from the group of normal hydrocarbons (up to C_{29}), the isoalkanes (present in the Nonesuch Shale), the anteiso alkanes,

the cyclohexyl alkanes (with a six-carbon ring at the end), the isoprenoids, including the C_{21}, and the polycyclic isoprenoids, the steranes
(with four rings), the pentacyclic triterpanes (with five rings), and some
hexacyclic hydrocarbons. These are a rather specific group of compounds, and it seems that such specific patterns of alkanes provide
persuasive evidence of their biological origin, in view of the large number
of theoretically possible, energetically almost equivalent, structures that
are *not* generally found. For example, the number of ways by which one

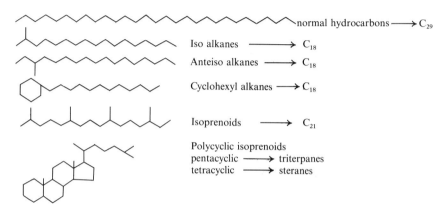

normal hydrocarbons ⟶ C_{29}

Iso alkanes ⟶ C_{18}

Anteiso alkanes ⟶ C_{18}

Cyclohexyl alkanes ⟶ C_{18}

Isoprenoids ⟶ C_{21}

Polycyclic isoprenoids
pentacyclic ⟶ triterpanes
tetracyclic ⟶ steranes

FIG. 4.18. Hydrocarbons found in rocks ranging back to the Archaean (3×10^9 years).

can construct a hydrocarbon containing five carbons is limited. There
are only three structures, not including rings, for C_5H_{12}: the normal
hydrocarbon, a form with a branched chain, and a third in which the
quaternary carbon atom is the centre. If there were some random way
by which methane were dehydrogenated to give this group of pentanes,
there should be a mixture of them, depending on how they were made
(at what temperature, whether or not they were in thermal equilibrium,
etc.), and all three should be present. Going on to higher hydrocarbons,
the number of isomers (possible structures with identical atomic compositions), becomes really large. For $C_{10}H_{12}$ there are 75 isomers, and
the energy difference between many of them is quite trivial; all the
isomers should be present if they were indeed formed in a random
manner. For $C_{20}H_{42}$ the number of isomers is 366 319, i.e. there are
366 318 possible arrangements other than phytane for $C_{20}H_{42}$. There are,
however, only two or three isomers usually found in nature. The degree
of specificity is the heart of the matter and has really persuaded most
investigators that these patterns do indeed represent the residues of

biological activity in the ancient sediments and could not have been arrived at in any abiological way.[28,29]

We should then try to make hydrocarbons experimentally by non-biological methods and see what they turn out to be. The first few experiments of this kind have been discussed above. Fig. 4.19 shows the gas-chromatographic analysis of the hydrocarbon fraction produced by hydrochloric acid on iron carbide. There is no series of discrete end-products: almost all the hydrocarbons appear.[30] This is one way of

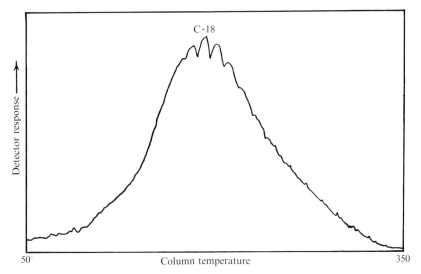

FIG. 4.19. Gas chromatogram of hydrocarbon fraction produced by hydrochloric acid on iron carbide.[22]

making hydrocarbons, and it appears to give clear support for the idea that the specificity we have seen among the hydrocarbons all the way back into the ancient rocks confirms the preliminary notion that they are of biological origin.[31] The methane-spark discharge method is another abio-logical method of producing higher hydrocarbons. This, again, shows a relatively smooth distribution (Fig. 4.20).[32] There exist in nature a number of samples that do not show a discrete distribution of hydrocarbons, and these are the ones we should be seeking as the abiogenic materials. The Trinidad Lake asphalt, shown in Fig. 4.20, which gives the saturated alkanes chromatogram, also shows a fairly smooth distribution, unlike the Posidonian Shale aliphatic hydrocarbons, shown in Fig. 4.20. The general pattern of smooth distribution in the Trinidad Lake hydro-carbons shows that a large variety of molecules are present even in the Trinidad Lake asphalt. This asphalt surely has in it some biological material that has drifted into it on the wind and circulates in it, but I do not know how deep the biological material is nor what its age might be.

There are at least two sites in which hydrocarbons have been found under geological conditions that tend to suggest that they are nonbiological in origin.[33] The Trinidad Lake asphalt is one, and near Leicester

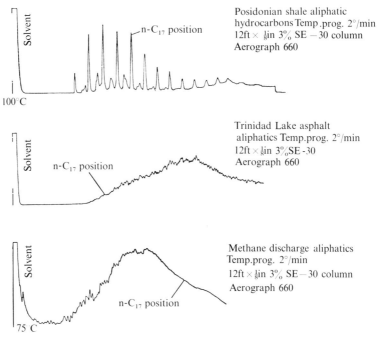

FIG. 4.20. Comparison of gas chromatogram of aliphatic hydrocarbons from Trinidad Lake asphalt with hydrocarbons from Posidonian marine shale and hydrocarbons synthesized by a spark discharge on methane.[32]

there is another, the Mount Sorrel formation. Fig. 4.21 (c) shows the gas-chromatographic analysis (on a different column from the one used to produce the curves of Fig. 4.20) of the hexane fractions of the Mount Sorrel sample. (The Posidonian shale in Fig. 4.21 (a) will serve as a reminder of the hydrocarbon pattern distribution in a typical biogenic shale.) Fig. 4.21 (b) is the chromatogram of the methane discharge products. The Mount Sorrel hexane fraction shows a multitudinous mixture of hydrocarbons, as in the methane-spark discharge. Thus, we now have two synthetic materials, the methane-discharge and the iron carbide hydrocarbons, that show the almost statistical distribution of hydrocarbons to be expected from a non-selective synthesis, and two natural occurrences that show a similar smooth curve.

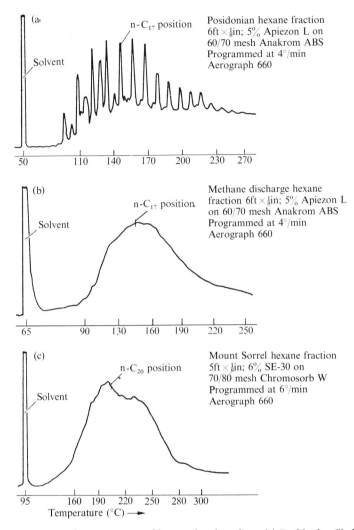

FIG. 4.21. Gas chromatograms of hexane fractions from (a) Posidonian Shale;
(b) methane discharge; (c) Mount Sorrel formation.[33]

We also had available another alleged abiogenic hydrocarbon in the
form of thucholite, a crystalline pegmatitic carbon from Sudbury,
Canada, which has been extensively described by Spence.[34] The thucholite
contains 20 per cent hydrocarbons, and its gas chromatogram is shown
in Fig. 4.22 (a). This is a peculiar 'abiogenic' gas chromatogram. Here
again there are few small peaks ahead of the n-C_{17} and n-C_{18} peaks
(presumably phytane and pristane). Fig. 4.22 (b), for the Soudan Shale,
shows a typical pattern of what is believed to be a biologically originated

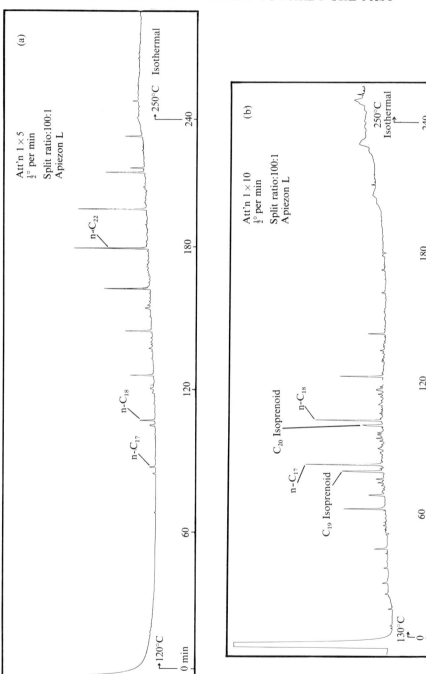

Fig. 4.22. (a) Gas chromatogram of hydrocarbons from thucholite from Sudbury, Ontario, allegedly abiogenic. (b) Gas chromatogram of total alkane fraction from Soudan Shale for comparison.

hydrocarbon. The resemblance between the thucholite and the Soudan chromatograms is striking.[35]

This remarkable similarity introduces a germ of doubt about the abiological significance of these discrete patterns. If we assume that the hydrocarbons from the Soudan Shale and the Fig Tree Shale are indeed biological residues, then the whole complex of enzyme systems that give rise to them must have been generated in the relatively short time-interval between 4700 million years and 3100 million years (the age of the Fig Tree Shale), giving only 1500 million years from the formation of the primeval earth to the presence of complex living organisms. While we do not yet have a very substantial estimate of the possible rate of chemical evolution that could have brought us to that point, the time available for this evolutionary period (1500 million years) has seemed short enough to some (particularly Sir Robert Robinson) to induce the suggestion that the germs of life were coeval (co-aggregated) with the earth itself. This is not a revival of the panspermia hypothesis of the last century, but rather a different notion. Sir Robert has said this several times.[36,37] His idea is that while the earth aggregated from cold dust, directly with that dust came the elements of microbiology. Sir Robert has argued that even the more recent oils, some of which we have examined, contain in them many abiological hydrocarbons; he thinks that, as the earth aggregated, these micro-organisms that came with the primeval dust used the abiogenic hydrocarbons on which to grow. This basic notion needs further experimental substantiation, of course.

A different suggestion comes from another set of experiments designed to determine the nature of experimentally produced hydrocarbon mixtures. This is a continuation of the experimentally produced hydrocarbon search, of which we have seen two examples, namely, the iron carbide and the spark discharge. Fig. 4.23 shows the gas chromatogram of the 'total' alkane fraction from a Fischer–Tropsch product.[38] A Fischer–Tropsch reaction is a reaction of carbon monoxide with hydrogen on a suitable catalyst (metal or metal oxide) to give hydrocarbons. In the Fischer–Tropsch alkanes we see a discrete pattern of hydrocarbons produced by a non-biological route:

$$CO + H_2 \xrightarrow{\text{catalyst}} \text{hydrocarbons.}$$

Fig. 4.23 (b) is a chromatogram of the total alkanes from the methane-spark discharge method, which shows a crude oil with a dominantly smooth distribution.

These observations have planted a new idea: that it is possible to generate discrete unbranched chains by some non-biological method. In fact, one more method was devised by Wilson in New Zealand.[39] This involves the addition of methyl groups made from the gas by radiation,

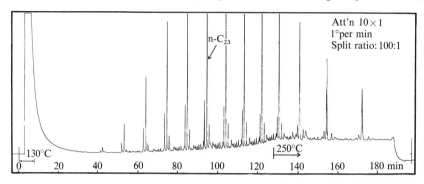

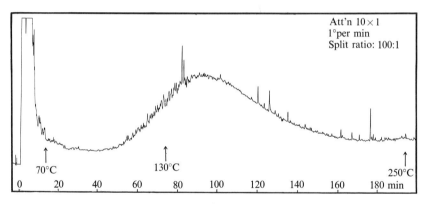

Fig. 4.23. (a) Gas chromatogram of 'total' alkane fraction from a Fischer–Tropsch reaction. (b) Gas chromatogram of total alkanes from methane-spark discharge.

or electrical discharge, or chemically to an oriented chain on the surface of water, or on the surface of a solid (Fig. 4.24). In the compressed film the methyl groups formed by ionization in the gas phase can reach only the tails of the hydrocarbons, because they are compressed. The results of this approach are shown in Fig. 4.25. In an experiment at 0 °C starting with a film of n-C_{16} acid, mostly normal C_{18} acid appeared after a certain amount of ionization, with very little of the iso compounds. At −10 °C the chain grew to C_{19} with a very small amount of the branched chain appearing. The basic idea is thus correct, and one can build up a linear unbranched chain in this very interesting non-biological way.[40]

The whole issue of possible specific abiogenic synthesis of certain families of compounds, even within the class of hydrocarbons, was thus opened up again.[41] With this idea of the possible abiogenic origin of the hydrocarbons and molecular markers we have hitherto regarded as solely biogenic, we shall begin the second part of our study of evolution

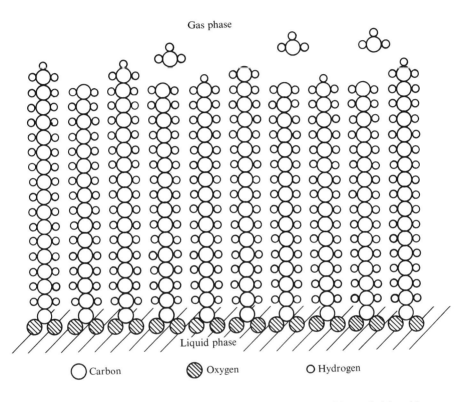

Gas phase

Liquid phase

○ Carbon   ◉ Oxygen   ○ Hydrogen

FIG. 4.24. Proposed mechanism for reaction of methyl radicals with a palmitic acid monolayer on water.[39]

130995

—the View from the Past. Up until now, we have taken the View toward the Past. We shall now undertake to discuss chemical evolution, beginning with the primeval earth, which will be the View from the Past, and see how far toward the present we can come, either in the laboratory or by observation of the stars and the planets. Having done that, we shall be able to build the bridge between the historical search we have just completed and the generational search we are about to begin.

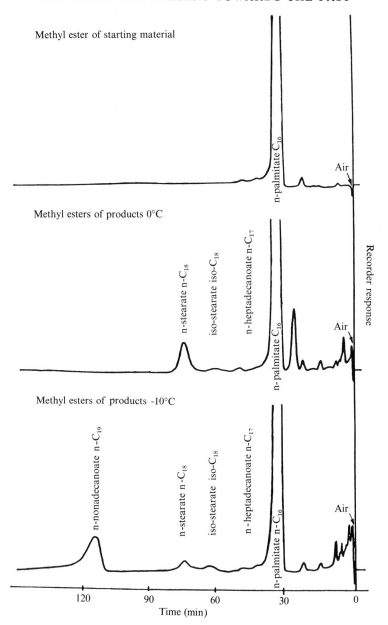

FIG. 4.25. Gas chromatograms of reaction-products from ionization of n-C_{16} fatty acid monolayer film, as in Fig. 4.24.[39]

REFERENCES

1. WHITE, W. S., and WRIGHT, J. C., The White Pine copper deposit, Ontonagon County, Michigan. *Econ. Geol.* **49**, 675 (1964).
2. EGLINTON, G., SCOTT, P. M., BELSKY, T., BURLINGAME, A. L., and CALVIN, M., Hydrocarbons of biological origin from a one-billion-year-old sediment. *Science* **145**, 263 (1964).
3. —— —— —— —— RICHTER, W., and CALVIN M., Occurrence of isoprenoid alkanes in a Precambrian sediment. *Advances in organic geochemistry*, 1964 (editors G. D. Hobson and M. C. Louis), pp. 41–74. Pergamon Press, London (1966).
4. CLOUD, P. E., JR., Significance of Gunflint (Precambrian) microflora. *Science* **148**, 27 (1965).
5. BELSKY, T., Organic geochemistry and chemical evolution. Ph.D. Thesis, University of California, Berkeley, December 1965. *University of California Lawrence Radiation Laboratory Report UCRL*-16566 (1966).
6. ORÓ, J., NOONER, D. W., ZLATKIS, A., WIKSTRÖM, S. A., and BARGHOORN, E. S., Hydrocarbons of biological origin in sediments about two billion years old. *Science* **148**, 77 (1965).
7. VAN HOEVEN, W., MAXWELL, JAMES R., and CALVIN, MELVIN, Fatty acids and hydrocarbons as evidence of life processes in ancient sediments and crude oils. *Geochim. Cosmochim. Acta*, in press.
8. CLOUD, P. E., JR., GRUNER, J. W., and HAGEN, H., Carbonaceous rocks of the Soudan iron formation (early Precambrian). *Science* **143**, 1713 (1965).
9. BELSKY, T., JOHNS, R. B., McCARTHY, E. D., BURLINGAME, A. L., RICHTER, W., and CALVIN, MELVIN, Evidence of life processes in a sediment two and a half billion years old. *Nature, Lond.* **206**, 446 (1965).
10. JOHNS, R. B., BELSKY, T., McCARTHY, E. D., BURLINGAME, A. L., HAUG, PAT, SCHNOES, H. K., RICHTER, W., and CALVIN, M., The organic geochemistry of ancient sediments. Part II. *Geochim. Cosmochim. Acta* **30**, 1191 (1966).
11. McCARTHY, E. D., and CALVIN, MELVIN. The isolation and identification of the C_{17} saturated isoprenoid hydrocarbon 2,6,10-trimethyltetradecane from a Devonian Shale: the role of squalane as a possible precursor. *Tetrahedron* **23**, 2609 (1967).
12. —— A treatise on organic geochemistry. Ph.D. Thesis, University of California, Berkeley, August 1967. *University of California Lawrence Radiation Laboratory Report UCRL* 17758 (1967).
13. ERICKSON, R. E., SHUNK, C. H., TRENNER, N. R., ARISON, B. H., and FOLKERS, K., Coenzyme Q: the structure of solanosol. *J. Am. chem. Soc.* **81**, 4999 (1959).
14. KOFLER, M., LANGEMANN, A., RÜEGG, R., GLOR, U., SCHIVIETER, U., WÜRSCH, J., WISS, O., and ISLER, O., Struktur und Partialsynthese des pflanzenlichen Chinons mit isoprenoider Seitenkette. *Helv. chim. Acta* **42**, 2252 (1959).
15. KARRER, P., and JÜCKER, E., *Carotenoids*, Elsevier, Amsterdam, London, and New York (1950).
16. THORNE, K. J. J., and KODICEK, E., The structure of bactoprenol, a lipid formed by *Lactobacilli* from mevalonic acid. *Biochem. J.* **99**, 123 (1966).
17. McCARTHY, E. D., VAN HOEVEN, W., and CALVIN, MELVIN, The synthesis of standards in the characterization of a C_{21} isoprenoid alkane isolated from Precambrian sediments. *Tetrahedron Lett.* #45, 4437 (1967).
18. HAN, JERRY, Ph.D. Thesis, University of California, Berkeley.

19. SCHOPF, J. W., and BARGHOORN, E. S., Alga-like fossils from the Early Pre-
cambrian of South Africa. *Science* **156**, 508 (1967).
20. HOERING, T. C., Criteria of suitable rocks in Precambrian organic geochemistry.
Carnegie Institution Yearbook, Vol. 65, p. 368 (1966).
21. ENGEL, A. E. J., NAGY, B., NAGY, L. A., ENGEL, C. G., KREMP, C. W. W.,
and DREW, C. M., Alga-like forms in Onverwacht series, South Africa:
Oldest recognized lifelike forms on Earth. *Science*, **161**, 1005 (1968).
22. MACLEOD, W. D., Jr. Combined gas chromatography–mass spectrometry of
complex hydrocarbon trace residues in sediments. *J. Chromatog.*, in press.
23. ABELSON, P. H., and HOERING, T. C., Carbon isotope fractionation in formation
of amino acids by photosynthetic organisms. *Proc. natn. Acad. Sci. U.S.A.*
47, 623 (1961).
24. PARK, R., and EPSTEIN, S., Carbon isotope fractionation during photosynthesis.
Geochim. Cosmochim. Acta **21**, 110 (1966).
25. DEGENS, E. T., *Geochemistry of sediments*, p. 282. Prentice-Hall, Englewood
Cliffs, New Jersey (1965).
26. HOERING, T. C., Reference 20, p. 369.
27. EGLINTON, G., and CALVIN, M., Chemical fossils. *Scient. Am.* **216**, No. 1, 32
(1967).
28. HOERING, T. C., The organic geochemistry of Precambrian rocks. *Research in
organic geochemistry* (edited by P. H. Abelson), vol. 2, p. 87. Wiley, New
York (1967).
29. MEINSCHEIN, W. G., Soudan formation: organic extracts of Early Precambrian
rocks. *Science* **150**, 601 (1965).
30. HOERING, T. C., Reference 20, p. 365.
31. MCCARTHY, E. D., and CALVIN, MELVIN, Organic geochemical studies. I.
Molecular criteria for hydrocarbon genesis. *Nature, Lond.* **216**, 642 (1967).
32. PONNAMPERUMA, C., and PERING, K. L., Aliphatic and alicyclic hydrocarbons
isolated from Trinidad Lake asphalt. *Geochim. Cosmochim. Acta* **31**, 1350
(1967).
33. —— —— Possible abiogenic origin of some naturally occurring hydrocarbons.
Nature, Lond. **209**, 979 (1966).
34. SPENCE, H. S., Remarkable occurrence of thucholite and oil in a pegmatite
dyke, Parry Sound District, Ontario. *Am. Miner.* **15**, 499 (1930).
35. MCCARTHY, E. D., Reference 12, p. 183.
36. ROBINSON, SIR ROBERT, Origins of petroleum. *Proceedings of the Royal Institution*, 1966, p. 110.
37. —— The origins of petroleum. *Nature, Lond.* **212**, 1291 (1966); Origins of oil:
a correction and further comment. Ibid. **214**, 263 (1967).
38. BELSKY, T., Reference 5, p. 92.
39. JOHNSON, C. B., and WILSON, A. T., A possible mechanism for the extra-
terrestrial synthesis of straight-chain hydrocarbon. *Nature, Lond.* **204**, 181
(1964).
40. MYERS, P. S., and WATSON, K. M., Principles of reactor design. #2. Pyrolysis
of propane. *Natn. Petrol. News* **38**, R-388 (1946).
41. STUDIER, M. H., HAYATSU, R., and ANDERS, EDWARD, Origin of organic matter
in the early solar system. I. Hydrocarbons. *Geochim. Cosmochim. Acta* **32**,
151 (1968).

PART II

THE VIEW FROM THE PAST TOWARDS THE PRESENT

5

CHEMICAL EVOLUTION

In our search for the residues of life we have found in the very oldest rocks, 3100 million years of age, the most stable kinds of materials that one could expect to find as biological residues, the simple straight-chain hydrocarbons, in which carbon atoms were strung, unbranched, in an orderly way. In addition we found the isoprenoid hydrocarbons, in which there was a single carbon atom branch on every fourth carbon atom of the long chain. Because of this very regular pattern of distribution, our first presumption was that these materials did indeed represent the products of biological activity. Towards the end of the previous chapter, however, some doubts were cast upon the biological significance of the residues in the ancient rocks. I shall now review briefly and expand somewhat on the kind of synthesis—some of the experiments, that is— in which these particular materials made their appearance.

I was once passing through an analytical laboratory in a large petro-chemical complex and, having seen one of the analytical sheets, noted that on it there was a very large component in the C_5 fraction— i.e. the material that had come off the distillation column containing compounds within the region of five carbon atoms, of the general type that could produce polyisoprenoids. In fact there was a large amount of isoprene itself, as well as of some other closely related materials. Further investigation revealed that it was common knowledge in the petrochemical industry that one could, by heating a mixture of two- and three-carbon compounds (ethane and propane, for example), or even two-carbon compounds alone, construct larger molecules from some smaller ones. Table 5.1 gives the results from a 1 : 1 mixture of two-carbon and three-carbon chains,[1] at a temperature of 1100 °K in the cracking furnace, with a contact time of approximately 1 s. Note the kinds of material that result. There is evidence not only of cracking but of an appreciable construction job as well. These four- and five-carbon pieces are not insignificant in amount. From either the pentadiene or isoprene we can obtain the polyisoprenoid skeleton by a 1,4 polymerization. This is a relatively large amount of specific material for what appears to be a random process, and a little further investigation showed what was happening. The relatively stable materials that could be formed were indeed formed; at such a high temperature and short time-interval, a total equilibrium among all the possible components was not obtained, but only a partial equilibrium. The essential feature, for this kind of construction of the polyisoprenoid structure, is a short time and a high temperature. In order to put two molecules together a high pressure would also be favourable, but this particular set of data was not obtained at very high pressure; it was a simple passage of gas through a hot tube at approximately 1 atmosphere pressure.

The construction of long linear chains is, we know, quite readily achieved with suitable catalysts. The unsaturated type of molecule, such as ethylene, is opened up and, in an excess of hydrogen, the linear hydrocarbons result. The same type of experiment can be done with a conjugate system (alternating double and single bonds), to create the isoprenoids. The long chains are in some cases produced in a rather specific configuration, according to the nature of the catalyst.

$$C_2H_4 \xrightarrow{\hspace{3cm}} \text{linear hydrocarbons } X(H_2C \cdot CH_2)_n \ldots Y$$

ethylene

$$\underset{\text{isoprene}}{CH_2 : \overset{\overset{\displaystyle CH_3}{|}}{C} \cdot CH : CH_2} \xrightarrow{\hspace{2cm}}$$

$$\left. \begin{array}{c} \\ \\ \\ \end{array} \right\} \text{isoprenoids}$$

$$\underset{\text{pentadiene}}{CH_2 : CH \cdot CH : CH \cdot CH_3} \xrightarrow{\hspace{2cm}}$$

$$X \left(\underset{CH_2 \cdot \underset{|}{C} : CH_2}{\overset{CH_3}{}} \right)_n \ldots Y$$

$$X \left(\underset{CH_2 \cdot CH : \overset{|}{C}}{\overset{CH_3}{}} \right)_n \ldots Y$$

If the reaction is carried out in excess hydrogen, X and Y will be H; if in water, H and OH.

Introduction : chemical evolution

This whole pattern reintroduces the idea that *modern biological molecules may have had abiological origins in the past*. In fact, this basic idea was

TABLE 5.1

Hydrocarbons synthesized in cracking furnace at 1100 °K from 1:1 mixture of two- and three-carbon chains

Product	Amount (per cent)
C_2H_4	30
C_3H_6	6–8
Butadiene (C_4H_6)	2
Pentadiene (C_5H_8)	~ 0.2
Isoprene	~ 0.2
Benzene (C_6H_6)	2
(H_2, CH_4, C_2H_6)	~ 60

promulgated in its present incarnation by J. B. S. Haldane in 1928[2] and by A. I. Oparin[3] at about the same time. The idea was presented only as a suggestion by these two men, and very little was done experimentally at that time. It was not until some twenty to thirty years later, just after the Second World War, that the experiments were begun. In 1950 one set of experiments was done in Berkeley (in our laboratory);[4] and in 1967 there were fifteen to twenty laboratories throughout the world devising experimental tests in which biologically important materials—materials that we had hitherto recognized as the products of complex and highly organized biological systems—were being made by abiological methods. The various laboratories that come to mind, all of which have published in 1967, are (apart from our laboratory in Berkeley):

University of California, San Diego (formerly University of Chicago)	Miller, Urey
University of Moscow	Paysinski, Oparin
University of Allahabad	Bahadur
University of Miami	Fox, Harada
Pennsylvania State University	Steinman
University of Bonn	Groth
University of Houston	Oró
Carnegie Institute of Washington	Abelson, Hoering
Ames Research Center, NASA	Ponnamperuma
Cambridge University	Markham
Mülheim	Schenck
Monsanto Co., St. Louis	Matthews
Salk Institute	Orgel

It is apparent that the people doing work of this kind are scattered all over the world, and experiments from most of these laboratories will be discussed below.

[This brings us back to the basic assumption underlying this entire exercise, namely, that 'living systems' appeared on the surface of the earth as a result of the interaction of 'primary energy sources' with some set of 'primeval molecules' (that is, non-living molecules) present on the surface of the earth.] There is, of course, an alternative idea, most recently enunciated by Sir Robert Robinson (see p. 95): that the 'living system' (whatever that was) arrived with the aggregation of the substance of the earth itself; that life, as we understand it, is coeval with matter throughout the universe. The problem, then, would be quite a different one and, at present, outside our capability.

To proceed along our primary path, we must therefore have some concept of the two factors involved in devising such an experimental, testable system. On the one hand is the *initial state* (starting-point) with which the whole evolutionary process is to commence; this would involve knowledge of the chemical composition of the primeval earth and the energy sources that affect it. On the other hand, we should also have to know something about the *final state* of the system (really the present state of the system, or the state towards which we expect the chemical system to move): namely, the qualities of living systems, so as to recognize them if and when they should appear in the experimental arrangements. I should like to take the second point, the question of the *final state*, first.

The 'final' state—life

At this point I do not seek to define a 'living system' in abstract terms that might be applicable to any collection of matter: for example, for use in extraterrestrial exploration, or perhaps in defining the nature of man. I shall defer these discussions until we have explored the specific problem of terrestrial life as we know it, namely, the elements of which it is constructed and the manner in which these elements are organized and function.

That function in a first approximation seems to be the directed use of energy to create order from a disordered, or less ordered, environment: in biological terms, growth and differentiation. We could also use chemical or thermodynamical terms, such as the principle in irreversible thermodynamics calling for a 'minimum rate of entropy increase', which might have a requirement for a maximum 'biomass' as a corollary. But I think this would lead to some confusion, and I shall limit myself to the more familiar biological terms. Further, function seems to be to generate and transmit the 'programme' for growth and differentiation to another system, that is, reproduction. Finally, function includes change in the 'programme' in response to a changing environment; the correlative biological terms would be mutation and selection.

The general types of materials of construction (shown in Table 5.2) are fairly well known. There are four major biopolymers: the proteins, nucleic acids, polysaccharides, and lipids. In addition there is a wide variety of small molecules that function in energy manipulations (flavins, chlorophylls, haem, ATP, coenzyme Q, etc.) and in material manipulation (coenzyme A, vitamin B_{12}), and there are also signal transmission

TABLE 5.2

General types of materials from which living matter is constructed

Atom	Molecule		Polymer	Cell
Hydrogen	Acid	$\rightarrow$	Lipid	Ribosome
Carbon	Sugar ⎞	$\rightarrow$	Cellulose, starch, etc.	Mitochondrion
Oxygen	Base ⎠	$\rightarrow$	Nucleic acid	Quantasome
				Nucleus
Nitrogen	Amino acid	$\rightarrow$	Protein	Membrane

materials (hormones, pheromones), and the like. These materials are organized into highly ordered structures at *all* levels, starting with the simple molecules. The patterns of the cellular and the sub-cellular structures and the details of the substrate structures are already well known. The polymer molecules, including enzymes, are organized into the ribosomes and mitochondria, chromosomes, quantasomes, etc. From this point up to the higher organelles of the cell (cytoplasmic structures, nuclei, membranes), the organization can be examined by the aid of such instruments as the electron microscope. I am assuming, then, that the materials of which the living cell is composed, and the level of organization that will be recognized, are known to the reader. We shall review the question of self-organizing systems later.

The initial state

The beginning is the initial state mentioned above, namely, the kind of primitive, or primeval, molecules that one might expect to find on the surface of the earth before any living material had arrived here—by whatever route. In order to learn more about the 'beginning' we must discuss briefly current ideas about the way in which the earth was formed and its relation to the solar system. This, in turn, requires that we get some concept of how the solar system itself was formed, and, finally, how the stars were formed.[5] The formation of stars and planetary systems is thus our next problem. The simple sequence of aggregation of dust and interstellar gas (hydrogen) is now the commonly accepted route to the formation of stars and planetary systems. This cold aggregation of dust and interstellar gas takes place with the conservation of angular momentum. The diffuse mass, containing many rotating bodies, has some net resultant angular momentum. As the diffuse mass aggregates, the angular momentum, which was originally distributed throughout the mass of gas and dust, must be retained, and a large, massive, slowly rotating, central core appears, heated by gravitational contraction. We thus have a sun, with smaller concentrations of matter circulating in larger orbits around the main mass in more rapid motion to conserve the angular momentum. This, of course, is a planetary system. Fig. 5.1 illustrates such a sequence of events.

The initial diffuse mass of gas and dust (indicated by condition (1) in Fig. 5.1) gradually aggregates by gravitational attraction.[5] The slowly rotating mass has to conserve angular momentum during this process, thus eventually giving rise to the planetary system. If the central mass (the sun) in this initial aggregation is more than 1·44 times that of our sun, it is unstable and will ultimately collapse because of gravitational attraction, with the temperature rising higher and higher until an internal thermonuclear reaction takes place in which the whole mass explodes. During that explosion many heavier elements are created from the initial hydrogen, and this is now redistributed; the whole process can then repeat itself until such time as the central massive star falls in its total mass below this critical limit (1·44 solar masses), in which case we have a lifetime for such a system of the order of 5 thousand million years (10^9 years), corresponding to the apparent lifetime of the sun, and providing a time-interval for the process of chemical to biological evolution. Terry and Tucker have made calculations about the frequency of such occurrences.[6] They point out that these occurrences within our galaxy occur roughly every 50 years; the supernovae—the large ones—occur every 50 million years. The purpose of this calculation was to try to understand the apparent discontinuities in the biological history, and perhaps even the geological history, of the earth. Terry and Tucker attributed some of the discontinuities in biological evolution to the appearance, once every 50 million years, of supernovae large enough to irradiate the surface of the earth with between 200 and 500 R, which is roughly half the lethal dose for humans.

A process such as that described above is the one more commonly believed to be that which gave rise to our present solar system about 5000 million years ago. Fig. 5.2 shows that solar system—the different kinds of objects it is composed of, and also their comparative size. Of

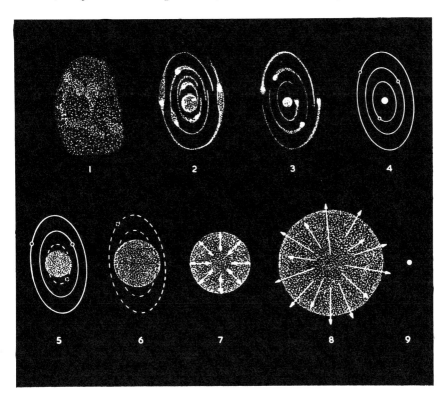

FIG. 5.1. Evolution of a star and a planetary system. Aggregation of interstellar dust by gravitational attraction forms a star and planets over a period of about 10 million years (1)–(4). The star enters the main sequence (4) and remains in equilibrium for about 8000 million years, gradually consuming hydrogen. The star then leaves the main sequence, expands into a red giant (5) and (6), and consumes its planets during the next 100 million years. After a few thousand years of pulsating as a variable star (7) it explodes as a supernova (8) and finally collapses as white dwarf (9).[5]

the nine planets, six have a total of thirty-one moons. The 1600 asteroids, which are scattered in the region between Mars and Jupiter, are fairly large bodies and there must be many thousands of much smaller ones. Note that the earth is third in the sequence and also note the relative sizes of the earth compared with its nearest neighbours, Mars and Venus, and compared with the giant planets, Jupiter and Saturn.

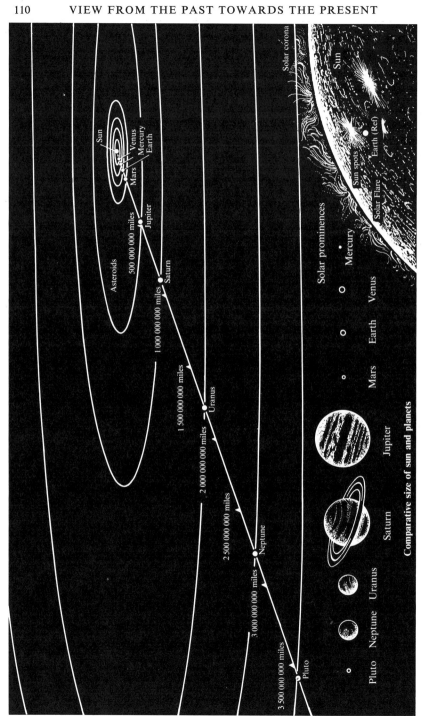

The solar system.[7]

With this point of view of the origin of planetary systems it has been possible to make reasonable estimates of the elementary composition of the primeval atmosphere of the earth.[8] Part of the evidence for the present view is, of course, the enormous dominance of elemental hydrogen in the cosmos. This hydrogen dominance shows up in the sun as well, and Table 5.3 shows the elemental composition of the solar material.

TABLE 5.3

Elemental composition of stellar material and the sun

	Stellar material (%)	Sun (%)
H	81·76	87·0
He	18·17	12·9
N, C, Mg	0·33	0·33
O	0·03	0·25
Si, S, Fe	0·01	0·004
Other elements	0·001	0·04

Our sun shows a somewhat larger component of heavier elements (0·044 per cent) than does the universe as a whole (0·011 per cent). This reflects its relative age as a star; the sun is a fairly 'middle-aged' star in the sense that there must have been a number of supernoval explosions whose materials have given rise to the heavier elements that are at present principal components of the earth.

The composition of the gaseous atmosphere (above the cloud deck) of the giant planets of our solar system again reflects the dominance of hydrogen and of the next-heavier elements, oxygen, nitrogen, and carbon. Table 5.4 shows the atmosphere of Jupiter (as it has been determined spectroscopically) above the cloud deck at a temperature of less than 120 °K.[9] The oxygen (as water) in the atmosphere of Jupiter is below the gaseous atmosphere at the low temperatures of Jupiter, and water is not therefore directly apparent in the atmosphere above the cloud deck.

The primitive earth, having a higher average temperature, is presumed to have had an atmosphere similar to this: mostly hydrogen with methane, water, and ammonia as its principal molecular constituents, before the gravitational loss of hydrogen. It is interesting to compare the elemental composition of the cosmos (stellar materials) with that of the present earth and of the life upon it. This is done in Table 5.5.[7] The most abundant elements are hydrogen and helium; next most plentiful are

TABLE 5.4

The atmosphere of Jupiter

Gas	(%)
CH_4	1
NH_3	0·05
H_2	60
He	36
Ne	5

TABLE 5.5

Relative abundance of elements

Element	Cosmos	Earth		Life	
		Atmosphere hydrosphere	Crust	Plant (%)	Animal (%)
Hydrogen	1000·0	2·0	0·03	10·0	10·0
Helium	140·0				
Oxygen	0·680	9·978	0·623	79·0	65·0
Carbon	0·300	0·0001	0·0005	3·0	18·0
Neon	0·280				
Nitrogen	0·091	0·003		0·28	3·0
Magnesium	0·029		0·018	0·08	0·05
Silicon	0·017		0·211	0·12	
Iron	0·008		0·019	0·02	0·004
Argon	0·004				
Sulphur	0·003	0·0005		0·01	0·25
Aluminium	0·0019		0·064		
Calcium	0·0017		0·019	0·12	2·0
Sodium	0·0017	0·0008	0·026	0·03	0·15
Nickel	0·0005				
Phosphorus	0·0003			0·05	1·0
Potassium	0·00008			0·32	0·35
Others	0·00015	0·011	0·020	0·04	0·156

carbon, nitrogen, silicon, and magnesium. Note that the living material in the biosphere on the surface of the earth is composed dominantly of hydrogen, oxygen, carbon, and nitrogen which, of course, is what one might have expected, in view of the nature of the elements themselves.

Such an initial terrestrial atmosphere as this may have been modified by a number of different processes. First, by the diffusional escape of an appreciable fraction of the initial atmosphere, the hydrogen and helium. These are the lightest elements and they would escape from the earth because its gravitational field is small in relation to those of the giant planets. The lack of neon and helium (particularly of helium) in the earth's present atmosphere speaks for such a process. In fact, a good deal of the first atmosphere of the earth must have been lost, and in a very short time; so the suggestion is that many of the primitive gases of the earth's atmosphere (with which we shall have to deal) were regenerated from inside the earth by the gradual heating-up of the crust, liberating quantities of these gases that had been trapped internally. Organic material of the same kind that appears in comets probably constituted the primitive earth's atmosphere: molecules with carbon–hydrogen bonds, carbon–nitrogen bonds, nitrogen–hydrogen bonds, and oxygen–hydrogen bonds. These would have been produced during the gravitational heating of the earth's interior, which would also produce hydrogen, methane, carbon monoxide, ammonia, water, etc. These are the materials with which most of the experiments (mentioned above) on the simulation of the primitive atmosphere have been done.

Such a primeval atmosphere would not remain static, especially when it receives so large an energy input. I have already mentioned the diffusional escape of hydrogen. In addition there will occur, because of the large energy input, the breaking of the initial carbon–hydrogen, nitrogen–hydrogen, and oxygen–hydrogen bonds, because the elements carbon, nitrogen, and oxygen are all in their totally reduced state with excess hydrogen (the principal element with which we start). This will be followed by reassembly of the fragments in some new combinations, perhaps with some loss of hydrogen gas which might be liberated by such a cracking operation.

Energy sources—inputs

In order to begin an examination of what could actually happen under these primitive earth conditions, we need to know a little about the kinds of energy input to which this atmosphere is likely to be subject. There are several approaches to this question: one is to say 'What is the energy input today?'; another is to try to extrapolate backwards to see if there is likely to be any difference between the energy input today and the energy input some 5000 million years ago. Consideration of the energy impinging on the earth's atmosphere today produces a list of at least five different sources, shown in Table 5.6. These sources are radioactivity (in terms of ^{40}K), ultraviolet light from the sun, volcanism, meteoritic impact, and lightning (electrical discharge). The problem of evaluating these energy sources is difficult, and in general entails non-equilibrium conditions, product quenching, and protection. Let us first consider ionization and the bond dissociation generally produced by high quantal energy inputs.

The value given in Table 5.6 for ^{40}K decay is estimated on the crust of the earth alone, using the present potassium distribution and its life-time, and we can extrapolate back to what it was 2600 million years ago.[10] The value for ultraviolet light is simply an estimate of the solar ultra-violet as it now impinges on the surface of the atmosphere. This is perhaps the best evaluation in the table, since it is one that can be measured without any estimate of what might, or might not, be present.

TABLE 5.6

Possible energy sources for primary chemical evolution

Source	Energy average over earth's surface (10^{20} cal/year)
Decay of ^{40}K (at present)	0·3
Decay of ^{40}K ($2·6 \times 10^9$ years ago)	1·2
Ultraviolet radiation shorter than 150 nm (1500 Å)	0·08
Ultraviolet radiation shorter than 200 nm (2000 Å)	4·5
Volcanism (from lava at 1000 °C)	0·04
Meteoritic impact	probably 0·05
Lightning	0·05

The volcanism is, again, a factor that is difficult to be sure of, because of the problem of estimating exactly how much material is brought to the surface each year, and at what temperature. The same can be said of meteoritic impact and electrical discharges; these are estimates only. The only estimates I have made myself are those for ultraviolet radiation and meteoritic impact. The latter had not been made before in these terms and context. We shall come back later to the source of information about meteoric impact, because this will constitute a very important contribution to the present discussion.

Ordinary slow thermal input, such as volcanism, would appear to be of only special application. By slow, I mean the contact of organic chemicals with hot rocks at 1000° or higher for extended periods of time. Obviously, such a process can lead to little except carbon. The initial separation and protection of products is really not solved by volcanic action alone. Under very special conditions, such as the right inter-mediate temperatures, which could lead to the concentration of the aqueous organic solutions to a dehydrated mixture, this energy input could be important.

There is one point I have assumed but have not yet discussed explicitly. It is quite clear that the very activity of these high-energy sources in taking apart the methane–ammonia–hydrogen–water atmosphere to create high-energy fragments for recombination into new configurations will eventually disintegrate the products of that rearrangement. Therefore, it is necessary to remove the product, in some way, from the field of action of these very high-energy sources if any progress is to be possible in this manner. This problem has been solved in four cases. For (1) ultraviolet light and (2) electrical discharge, most of which is absorbed in the high atmosphere, the gravitational separation of the product molecules is the principal method. Simple gravitation or aggregation and then precipitation in the form of rain enhances the gravitational rate of settling from the higher atmosphere, and actually moves the initial products, which are formed high up, out of the range of the ultraviolet at the top of the atmosphere and out of the electrical discharges in the upper reaches. The methods of (3) radioactivity and (4) volcanism would achieve separation by diffusion and convection in the sea, in which the volumes of the ocean in contact with the radioactivity and heat are continually in motion, thus removing the products from the direct action of the ionizing radiation or the heat, as the case may be.

It is obvious from Table 5.6 that ultraviolet light is far and away the greatest source of energy input. On the other hand, the crucial point is really the effectiveness of this kind of energy in producing changes in the molecules that will lead towards some chemical evolutionary process, not necessarily the total amount of energy. Suffice it to say that the amounts of energy in any one of these forms is ample if it is used in any effective way. In fact (as will become apparent as the discussion continues) all the four sources—radioactivity (or its equivalent), electrical discharge (lightning), ultraviolet light, and heat (long-term heat flow such as in volcanism)—have been tested in a wide variety of ways for the production of biologically important materials.

I have mentioned all the sources except meteoric impact explicitly as means that seem to provide ways of separation. What I want to add now is a discussion of meteoric impact, which has only recently become a useful notion in this kind of chemical evolutionary experimentation. It came about in my mind in two ways. At the beginning of this chapter I mentioned the commercial process for materials of the isoprene and polyisoprene type. This is a thermal cracking process in which the gases are passed through a hot tube for a short period of time. In other words, a pulse of heat is applied to the starting material. This looked so promising to me that I began to wonder how it could have occurred on the primitive earth, and I began to think in terms of fumaroles (holes and chimneys in the hot rocks with gases coming up through the chimneys) and things of that kind. This approach was not too satisfactory, and did not lead very far. A week or so later I was trying to devise ways in which this pulse of heat might have been applied to the primitive atmosphere, when a man came into my office in Berkeley for a discussion. He presented an idea to me, with the information that it had not been explored experimentally, and he felt it was important. This man was an engineer who had been working on space-vehicle re-entry. He told me how he happened to be doing that. You may remember that when the first *Mercury* space capsule with its man in it went out and was re-entering, we lost radio contact with it for a short period of time, causing great consternation. Apparently on re-entry, as was very quickly realized, the atmosphere is heated so much by the incoming vehicle that it ionizes the molecules around it, producing a conducting layer, and thus cutting off the radio contact for a short period of time. This phenomenon is true of any re-entering vehicle, and it became important to learn about the nature and shape of these ionization fields. During the course of that work he had come to realize the enormous amounts of energy that were being put into the earth's atmosphere in a very short time and in a very peculiar way. This led to some calculations. The important thing was that he thought in terms not only of the re-entry of a vehicle from space, but of what meteorites might do to a primitive atmosphere.[11,12] He recognized this possible application of his work and had actually tried to get an experiment done in which somebody fired a supersonic bullet in a cannon filled with methane–ammonia–water. This, I might say, is a rather expensive experiment to do. I think each firing of the cannon costs some thousands of dollars, and one does not get many chemical products from it. It is not a very efficient synthetic method! He was seeking some other application of this meteoritic impact notion when he came to my office.

This kind of idea introduces at least two new chemical environments for synthesis and separation. The first of these is the shock wave in the gas and the second is the final impact with the solid earth or liquid water (the sea). Fig. 5.3 (from Hochstim's paper) shows a diagram of hypersonic flow around a sphere in which is shown the 'stagnation region', just ahead of the boundary layer. These regions of the shock wave are defined in terms of the temperature and pressure in the stagnation region (P_s and T_s), the mass of material in the stagnation region just ahead of the meteorite (m_s), and the mass of the material in the inner wake (m_4R). The values Hochstim found are shown in Table 5.7.

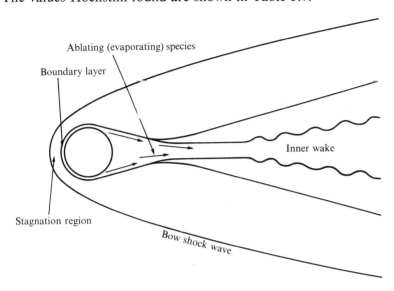

FIG. 5.3. Hypersonic flow around a sphere.[12]

TABLE 5.7

Hochstim's calculations (see Fig. 5.3)

Meteorite radius (m)	Meteorite velocity (km s^{-1})	Temperature in stagnation region, T_s (°K)	Pressure in stagnation region, P_s (atm)	Mass of material in stagnation region, m_s	Mass of material in inner wake $m_4 R$	Kinetic energy (ergs)
1	5	7000	200	3 kg	1·6 kg	$2·6 \times 10^{12}$
1	11	16 500	1500	3 kg	1·6 kg	$1·3 \times 10^{13}$
1	30	70 000	10 000	3 kg	1·6 kg	$9·4 \times 10^{19}$
500	11	16 500	1500	4×10^5 tons	2×10^5 tons	$1·6 \times 10^{27}$
500	30	70 000	10 000	4×10^5 tons	2×10^5 tons	$1·2 \times 10^{28}$

For a meteorite of 500-m radius coming in at 11 km/s, T_s is 16 500 °K, P_s is 1500 atm, and the mass of the material in the compressed wave is 4×10^5 tons, which is a rather large quantity. This type of chemistry can be carried out in a device that attempts to simulate the same conditions that are found ahead of the meteorite (Fig. 5.4). Instead of a meteorite coming into the atmosphere, we put a low-pressure atmosphere into a glass tube (the 4-ft section on the left of Fig. 5.4), and we put behind it the equivalent of the meteorite in the form of high-pressure gas (hydrogen) in a brass tube (the $7\frac{1}{4}$-in section on the right). The low-pressure gas in the glass represents the atmosphere (the pressure of the gas is 5 torr). The low-pressure glass chamber is separated from the high-pressure brass chamber by a thin diaphragm of aluminium foil. We increase the

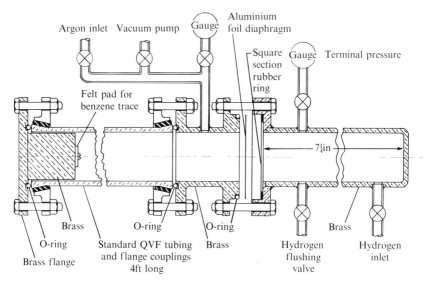

FIG. 5.4. Apparatus for compression-wave experiments.

gas pressure in the brass chamber until we reach the pressure at which the aluminium diaphragm will break; it then breaks very quickly, and at this instant a high-pressure wave of gas impinges on the low-pressure gas. This high-pressure front compresses the low-pressure gas, by a factor of roughly 5000, so rapidly that the low-pressure gas is heated up.[13] An incoming meteorite would affect the atmosphere in exactly the same way. The front of the high-pressure wave as it enters the low-pressure region might be conceived of as a *solid* front analogous to that of a meteorite surface entering the atmosphere. When the aluminium diaphragm bursts, a flash of light is seen as the gas is heated up. The flash is particularly visible at the far end, because the compression wave hits the end and bounces back; when the two waves double up, the temperature goes up to several thousand degrees, enough to excite the gas in the glass tube and make it emit light. The pressure is 120 lb/in² when the diaphragm breaks. With only argon in the glass tube there is not enough organic material present to show anything but the light flash. No perma-

PLATE 3

Soudan Shale (Michigan and Canadian Shield) ($2 \cdot 5 \times 10^9$ years)

PLATE 4

Fig Tree Chert (Swaziland) ($3 \cdot 1 \times 10^9$ years)

nent chemical change is apparent. To show this apparatus doing some 'chemistry' we have to put in the glass tube some organic material that can be changed by the compression wave. Relatively little of this kind of study has been done, and none at all using the 'primeval' gas mixture methane–ammonia–water–hydrogen. An estimate of the amount of chemical change one can expect has been made using dry air as the gas, but dry air is not a very useful combination for organic chemistry.[14]

Let us consider the frequency of such meteoritic energy input on the surface of the earth. There are at least 1600 asteroids visible in the asteroid belt between Mars and Jupiter. In order to be seen an asteroid must be the size of a small moon. One of these very large asteroids striking the surface of the earth would change the whole nature of that surface. Various people have made estimates of the number of large meteorites that might have struck the earth. These very interesting objects leave evidences on the surface of the earth (for example, the great meteorite crater in Arizona, and the result of the one that must have struck in Siberia in 1908). The estimate of the rate of infall of large meteorites, i.e. those big enough to penetrate the atmosphere and make a crater in the surface of the earth, is one every 10 000 to 20 000 years; and the rate was probably higher in the early history of the earth. It stands to reason that if the earth's orbit has not changed very much, it is gradually sweeping out these meteorites and there must be fewer of them now than there were 5000 million years ago. They have been collected by the earth, so to speak. The estimates of one every 10 000 to 20 000 years (for the largest meteorites now) suggest a much larger rate at the beginning of the earth's history.

A meteorite of 500-m radius striking the surface of the earth in normal dry air will expose 100 000 metric tons of air in one fall. This is not, from our point of view, a very useful gas combination: nitrogen, oxygen, with a small amount of CO_2. It will, however, produce 10^8 kg of NO, 10^7 kg of oxygen atoms (which do not last very long), and 10^6 kg of CO_2. So the amount of chemistry that this pulse of pressure and temperature can effect is very large, and the number of occurrences of this type is not insignificant—one every 10 000 to 20 000 years.

These events seem to provide all the requisite conditions that we described earlier: the shock wave in the atmosphere giving a temperature pulse of thousands of degrees, a pressure pulse of many atmospheres, and the final collision with the earth, or rather the sea, producing in the immediate vicinity of the newly made materials rain that will then move these newly made organic compounds out on to the surface of the earth.

The other four methods of energy input (radioactivity, electrical discharge, ultraviolet light, and heat) have also been explored and a wide variety of materials have been produced. We shall subsequently go over each of these methods in turn. The next stage of our investigation will be to examine the variety of small molecules produced by the impingement of these different kinds of energy: radioactivity, ultraviolet light, electrical discharge, and possibly meteoritic input through shock waves. After that we shall see if means can be devised for allowing the small molecules to 'grow' from larger ones by the various polymerization processes (vinyl polymerization,[15] condensation polymerization[16]) mentioned at the beginning of this chapter (p. 104).

REFERENCES

1. McCarthy, E. D. and Calvin, Melvin, Organic geochemical studies. I. Molecular criteria for hydrocarbon genesis. *Nature, Lond.* **216**, 642 (1967).
2. Haldane, J. B. S., The origin of life. *Rationalist Annual* (1929).
3. Oparin, A. I., Proiskhozdenie zhizny, *Izd. Moskovski Rabochii* (1924); Vozniknovenie zhizny na zembi, *Izv. Akad. Nauk SSSR* (1936); The origin of life (translation of Russian edition, 1936, editor S. Margulis). Macmillan, London (1938).
4. Garrison, W. M., Morrison, D. C., Hamilton, J. G., Benson, A. A., and Calvin, M., The reduction of carbon dioxide in aqueous solutions by ionizing radiation. *Science* **114**, 461 (1951).
5. Gade, C. M., *Other worlds than ours.* Museum Press, London (1967).
6. Terry, K. D. and Tucker, W. H., Biologic effects of supernovae. *Science* **159**, 421 (1968).
7. Newell, H. E., NASA's space science and applications program (a statement presented to Committee on Aeronautical and Space Sciences, U.S. Senate, 20 April 1967, p. 537). *NASA Publication* #*EP*-47 (1967).
8. Urey, H. C., *The planets.* University of Chicago Press (1952).
9. Sagan, C. E., Lippincott, E. R., Dayhoff, M. O., and Eck, R. V., Organic molecules and the coloration of Jupiter. *Nature, Lond.* **213**, 273 (1967).
10. Swallow, A. J., *Radiation chemistry of organic compounds*, p. 244. Pergamon Press, London (1960).
11. Gilvarry, J. J. and Hochstim, A. R., Possible role of meteorites in the origin of life. *Nature, Lond.* **197**, 624 (1963).
12. Hochstim, A. R., Hypersonic chemosynthesis and possible formation of organic compounds from impact of meteorites on water. *Proc. natn. Acad. Sci. U.S.A.* **50**, 200 (1963).
13. Pritchard, H. O., Shock waves. *Q. Rev. chem. Soc.* **14**, 16 (1960).
14. Green, E. F. and Toennies, J. P. *Chemical reactions in shock waves.* Academic Press, New York (1965).
15. Natta, G., Precisely constructed polymers. *Scient. Am.* **205**, No. 8, 33 (1961).
16. Tsuruta, T. and Inoue, S., Well-ordered polymers. *Sci. Technol.* No. 71, November 1967, p. 66.

6

PRIMITIVE (PREBIOTIC) CHEMISTRY

WE have examined a little of the history of the solar system, with a view to trying to determine what the beginning of our chemical evolutionary sequence might have been, and we came to the conclusion that the primeval atmosphere on the surface of the earth was, in all probability, a reducing one, containing essentially the four elements carbon, hydrogen, nitrogen, and oxygen, primarily in their reduced form, as the simplest of molecules.[1,2] There are some who believe that this heavily reduced atmosphere, dominated by hydrogen, which would have been the first of the earth's atmospheres, was very early lost, and that a secondary atmosphere was generated from the captured gases of the earth's crust. Under these circumstances the same elements would have been present in the atmosphere but not in such a thoroughly reduced form; there would have been some hydrogen; the nitrogen would have been perhaps more in the form of elemental nitrogen; the carbon instead of being mostly methane might have been more dominantly carbon monoxide. But, nevertheless, it would be a reduced atmosphere in spite of that. We also examined the varieties of energy inputs that could impinge upon this collection of simple molecules and begin to transform them.

We now begin our investigation of the evolution of 'small' molecules. It is useful to review the general sequence of the remaining chapters. After having examined (1) the primary evolution of the 'small' molecules we shall go on to discuss (2) the growth of molecules, i.e. the generation of polymeric materials, (3) the participation of the effect of catalytic function and the evolution of catalytic function and reflexive catalysis, (4) the problem of structural self-assembly of macromolecules, (5) the evolution of chemical systems—whole systems of transformations —followed by (6) a discussion of the last three stages, which are the concentration and isolation of these chemical systems into small compartments (membranes), and the appearance of a general structure of living cells as we now know them. We shall also discuss the applications of these notions to the exploration of the nearest neighbours of the earth, that is, the moon, Mars, and Venus, for evidences of organic and biological material. Lastly, we shall explore the possible human significance of the whole exercise.

TABLE 6.1

Schematic representation in chemical terms of the formations that have to be accomplished from the atoms to produce the structure of the cell

Atom	Molecule		Polymer		Cell
Hydrogen	Acid	$\rightarrow$	Lipid	⎱	⎧Molecular aggregation
Carbon	Sugar ⎱	$\rightarrow$	Cellulose, starch, etc.	⎰	+⎪
Oxygen	Base ⎰	$\rightarrow$	Nucleic acid	$\rightarrow$	⎨Autocatalysis
Nitrogen	Amino acid	$\rightarrow$	Protein	⎰	+⎪
					⎩Membrane assembly

Table 6.1 summarizes the general line of development from atom to molecule to polymer to cell, showing the four elements that are involved in the primeval atmosphere (carbon, hydrogen, oxygen, nitrogen), the simple molecules derived from them (acid, sugar, base, amino acid), the polymers that, in turn, are created from the simple molecules (lipids, cellulose, nucleic acids, proteins), and beyond these, higher-ordered structures, self-organizing systems (self-assembling systems) that will give rise to the limiting membrane and the catalytic system for their creation and preservation.

The evolution of 'small' molecules

The nature of the energy inputs discussed in Chapter 5 (radioactivity or its equivalent, ultraviolet light, electric discharge, heat) is known, and to a first approximation, all these energy inputs give the same general result: the same kind of collection of molecules is formed from the primeval atmosphere. What is obtained in the first approximation depends more upon the composition of the starting material than it does on the nature of the energy input.

Experiments with primitive atmospheres

The general plan of most experiments with primitive atmospheres is to devise an initial gas composition, and then to remember that changes are introduced by whatever energy input we allow to act on it, and that this energy input will ultimately degrade the product of the first action.[3,4] Some method of separating the product of the first action from the energy impingement has therefore to be included in the design of the experiment. We need (1) to devise the gas composition and a mode of separation of the initial product; (2) to select an energy input; (3) to allow the system to operate for some time, i.e. allow the energy input to impinge on the gas mixture and allow the separation process to take place; and (4) to perform a variety of analytical operations on either the residual gas, the solution, or the non-volatile residue.

Water	Carbon monoxide	Carbon dioxide	Methane	Hydrogen	Ammonia
$H-O$	$C \equiv O$	$O = C = O$	$H-C-H$	H	$N-H$

$$H-O \quad C \equiv O \quad O=C=O \quad H-\overset{\overset{\displaystyle H}{|}}{\underset{\underset{\displaystyle H}{|}}{C}}-H \quad \overset{\displaystyle H}{\underset{\displaystyle H}{|}} \quad \overset{\displaystyle H}{\underset{\displaystyle H}{|}}N-H$$

Water   Carbon monoxide   Carbon dioxide   Methane   Hydrogen   Ammonia

$$H-C \equiv N \quad HN(C \equiv N)_2 \quad H-\overset{\overset{\displaystyle O}{\|}}{C}-OH \quad H-\overset{\overset{\displaystyle H}{|}}{C}=O \quad HOCH_2-\overset{\overset{\displaystyle H}{|}}{C}=O \quad CH_3-\overset{\overset{\displaystyle O}{\|}}{C}-OH$$

Hydrocyanic acid   Dicyanamide   Formic acid   Formaldehyde   Glycolaldehyde   Acetic acid

$$HO-\overset{\overset{\displaystyle O}{\|}}{C}-CH_2-CH_2-\overset{\overset{\displaystyle O}{\|}}{C}-OH \quad H_2N-CH_2-\overset{\overset{\displaystyle O}{\|}}{C}-OH \quad CH_3-\overset{\overset{\displaystyle O}{\|}}{\underset{\underset{\displaystyle NH_2}{|}}{C}H}-\overset{\overset{\displaystyle O}{\|}}{C}-OH \quad HO-\overset{\overset{\displaystyle O}{\|}}{C}-CH_2-\overset{\underset{\underset{\displaystyle NH_2}{|}}{\displaystyle }}{C}H-\overset{\overset{\displaystyle O}{\|}}{C}-OH$$

Succinic acid   Glycine   Alanine   Aspartic acid

The general results of such experiments (and I shall describe three types in some detail) are shown in Fig. 6.1, which gives some idea of the kinds of products we find in our chemical evolutionary exploration. The primary or primitive gas mixture is represented by the top row, in which I have included both the fully reduced mixture and the partly reduced one. The second row represents some of the very first transformation products that result from energy inputs into molecules of the first row. The bottom row shows some of the first kinds of the somewhat larger groups of atoms that appear. The general pattern of these products includes almost all, or at least many of, the common metabolites used by *all* today's organisms, except, perhaps, for the most interesting of the group, hydrogen cyanide, HCN, which is the first molecule on the left of the second row in Fig. 6.1. One would hardly speak of hydrogen cyanide as being a common, everyday metabolite, and yet, as we shall see later on, it has assumed an importance (or seems to have assumed a kind of importance) that is rather unexpected for that kind of material.

One of the early experiments in the modern series of chemical evolutionary experiments was done in Berkeley in 1950.[5] It was the result of a curious concatenation of circumstances: a number of things happened to come together about that time, such as my own interest in the subject, the availability of labelled carbon (carbon-14), which enabled us to follow a carbon atom wherever it went in the products, and the availability of good energy sources, such as cyclotrons. This combination of circumstances produced the first experiments in the

TABLE 6.2

Results of primitive atmosphere experiment (1950, Berkeley)

Dose (μA He/h)	0·75	0·042
$p(CO_2)$ (mm)	2·4	2·9
$p(H_2)$ (terminal) (mm)	208	8
$C(Fe^{2+})$	0·8	—
Formic acid	4% of CO_2 (sol.)	22% of CO_2 (sol.)
Formaldehyde	0·1% CO_2 (sol.)	0·13% CO_2 (sol.)

modern sequence—in 1950. These experiments were, as a matter of fact, done with not very well-reduced atmospheres. We used CO_2, water, Fe^{2+} in solution, and hydrogen as the primary reducing agent, and the radiation was accomplished by 40-meV helium ions made in the 60-in cyclotron. This experiment was essentially one that simulated terrestrial crust radioactivity. The results are given in Table 6.2. The formic acid and formaldehyde are expressed in terms of the initial proportion of the carbon dioxide. In the second experiment a good fraction of the CO_2 was reduced. Since we were irradiating the water solution directly there was no provision for the continuous separation of the products; they would therefore be degraded by additional energy input. Unfortunately for us, two other things were left out of this experiment. First, there was no nitrogen in it, and therefore many of the more interesting compounds (referred to below) did not appear. Secondly, the atmosphere was not fully reducing but only slightly reducing, with a small amount of hydrogen and a catalytic amount of Fe^{2+}. We therefore missed the major step forward that was taken later by Miller. Successive experiments started with some of the products further down the chemical evolutionary line, including acetic acid, and other products resulted: succinic acid, lactic acid, malic acid, fumaric acid, etc.

(a) *Electric-discharge experiments*

Two years after the experiment just described was carried out, Stanley Miller in Urey's laboratory at the University of Chicago, following

Urey's analysis of what the primitive atmosphere should be (i.e. reducing and with ammonia in it), did the next crucial experiment.[6] This changed the pattern of the ongoing activity. Miller introduced ammonia and much more hydrogen into the gaseous atmosphere, and the carbon was in the first instance in the form of methane. He started the methane–ammonia–hydrogen–water system irradiations, which have since become very common. At about the same time (1955) one of the very few experiments using ultraviolet light of wavelengths shorter than 185 nm was performed on the methane–ammonia–hydrogen–water mixture upon which Miller had used electrical discharge; the primary products were urea, formic acid, and formaldehyde. Some oxidized atmospheres (CO_2 and NH_3) were used in the experiments, as well as CO and water; these experiments were done by Groth in Germany[7] and by Terenin in Russia.[8]

Let us now return to the generation of the second group of compounds (those in the bottom row of Fig. 6.1) from the primitive reducing atmosphere. I should like here to refer to a well-known reaction: the alkaline condensation of formaldehyde to create a whole series of carbohydrates. We performed the same type of experiment again with the formaldehyde from the earlier radiation, and we obtained a series of carbohydrates such as the ones shown in Fig. 6.2. This is the usual pattern of construction of carbohydrates, from formaldehyde to glycolaldehyde, etc. The alkaline condensation of formaldehyde is usually done on limestone, and shows the presence of C_2, C_3, C_4, C_5, and C_6 sugars. There is a restraint on this experiment because of the lability of the sugars themselves, and one has to provide some way of trapping them in more complex materials, thus removing the sugars from the reactive sites. This is one of the succeeding stages of chemical evolutionary experimentation that will be described below.

We shall see that a similar group of materials can be formed directly in the irradiation of methane–ammonia–water, probably produced by the ammonia basicity alone, from the primarily formed formaldehyde.

Miller's experiments were done in 1953. He used electric discharge instead of radioactivity, thus representing the third type of energy input.[9] The first kind of energy input was two kinds of radioactivity—heavy-particle (alpha) and light-particle (electron) bombardment. The second kind of energy input was ultraviolet radiation, which up till now has not been very important as an experimental tool. The third kind was the electric discharge, which constitutes the principal experimental method that has been applied in this kind of test ever since Miller himself used it. The apparatus used by Miller is shown in Fig. 6.3. It is very simple, providing for the two required qualities mentioned above. The discharge is in the gas bulb at the top right; the boiling water is in the pot at the bottom left; the steam rises and pushes the methane–ammonia–hydrogen mixture round the loop; the condensing water coming down through the condenser washes the product out into the trap and it collects in the pot.

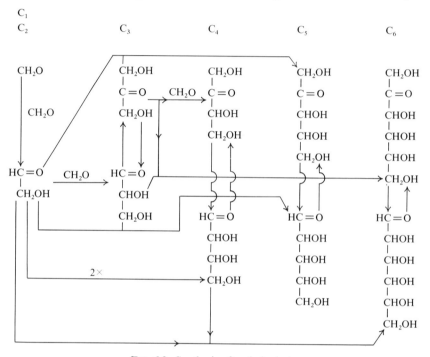

FIG. 6.2. Synthesis of carbohydrates.

Samples can be taken out of the trap at any time without opening the whole system. This apparatus provides both the energy input into the gas and a device for separating the product from the energy input; once the products are formed, as long as they are non-volatile, they will stay in the pot. This is the first of the simple systems that provide the ways of putting energy into the system and taking the products out of the energy source and running the apparatus for long periods of time. With this apparatus and initial chemical system a wide variety of products was obtained, and the field was thus opened up enormously. Fig. 6.4 shows some of the kinetic results. The ammonia disappears with time, and the concentration of the HCN rises with time.[10] The amino acid concentration also rises with time and as the amino acid concentration rises, the HCN concentration drops. Note that there is a group of aldehydes that remain in a fairly stationary state, indicating that they are being formed and used up; as the carbon runs out, the aldehydes disappear as well.

The principal products analysed in this series of experiments were the HCN on the one hand, and the aldehydes and amino acids on the other.[9]

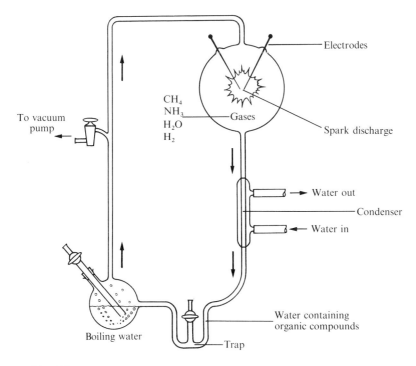

FIG. 6.3. Apparatus used by Miller for the synthesis of amino acids by electric discharge.[9]

The reason for this will soon become apparent. Before discussing the mechanism of these reactions it is perhaps worth while to discuss the content of the reaction-product mixture in somewhat more detail; this is shown in Table 6.3. A wide variety of amino acids are produced, dominated by glycine and glycolic acid and alanine and lactic acid, with a large amount of formic acid, which corresponds to our earlier observations. The parallelism of the two-carbon amino acids and hydroxy acids and the three-carbon amino acids and hydroxy acids led Miller to propose that these came about by the formation of formaldehyde from the primary ionization of the methane, followed by reaction with the water to give methanol, and a secondary ionization, eventually giving formaldehyde. The formaldehyde then reacted with the HCN and ammonia in a typical Strecker synthesis to give either the glycolic acid (hydroxy acid) or the corresponding amino acid. The nitrile groups, which upon hydrolysis give the carboxyl group, add first to the carbonium carbon, followed

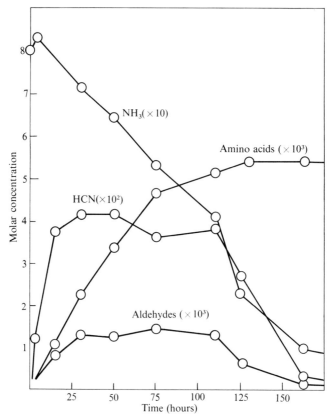

FIG. 6.4. Concentrations of ammonia, hydrogen cyanide, and aldehydes in the U-tube of Miller's apparatus and of amino acids in the small flask while sparking a mixture of methane, ammonia, water, and hydrogen.[10]

by reaction with water or ammonia, according to which of the two molecules is dominant, to give hydroxyl or amino compounds in the succeeding reaction. The same kind of reaction can be assumed for acetaldehyde, which would arise by the ionization of methane to give hydrogen atom and methyl radicals, the hydrogen atom abstracting a hydrogen atom from the formaldehyde and giving back H_2 and the formyl radical, which then could react further to form the acetaldehyde. These reactions are shown below:

TABLE 6.3

Yields from sparking a mixture of CH_4, NH_3, H_2O, *and* H_2;
710 mg of carbon added as CH_4

Compound	Yield moles ($\times 10^5$)
Glycine	63
Glycolic acid	56
Sarcosine	5
Alanine	34
Lactic acid	31
N-Methylalanine	1
α-Amino-n-butyric acid	5
α-Aminoisobutyric acid	0·1
α-Hydroxybutyric acid	5
β-Alanine	15
Succinic acid	4
Aspartic acid	0·4
Glutamic acid	0·6
Iminodiacetic acid	5·5
Iminoacetic-propionic acid	1·5
Formic acid	233
Acetic acid	15
Propionic acid	13
Urea	2·0
N-Methyl urea	1·5

(b) *Electron-bombardment experiments*

Following Miller's experiment we then introduced nitrogen, in the form of ammonia, into our reaction mixture together with excess hydrogen. Using methane as our carbon starting material we did the entire experiment again, and used electron bombardment as an energy input, simulating the ^{40}K of the earth's crust.[11] We obtained a much wider variety of materials, as shown in Table 6.4. There is a very large yield

TABLE 6.4

Identification of compounds from irradiation experiment from
^{14}C-labelled methane, ammonia, and water

Experiment M22 ^{14}CH$_4$, NH$_3$, H$_2$O, PH$_3$(NH$_4$PO$_3$)$_x$	
Acid fraction	45·5%
Basic fraction	17·44% separation on Dowex 1 and Dowex 50
Non-ionic fraction	21·3%
HCN	0·45% of total
Adenine	{ 0·203% of basic fraction { 0·034% of total
5-Aminoimidazole carboxamide	{ 0·105% of basic fraction { 0·018% of total
Lactic acid	{ 2·21% of acid fraction { 0·99% of total

The two unknown other dominating acids (not including lactic acid) account for 29% and 17·8% respectively of the acid fraction.

Glycine	{ 0·2% of basic fraction { 0·03% of total
α-Alanine	{ 1·07% of basic fraction { 0·18% of total
Aspartic acid	{ 0·2% of basic fraction { 0·03% of total

The two unknown (not urea or guanidine) dominating basic compounds (ninhydrin-positive) account for 20·9% and 11·9% of the basic fraction.

of HCN under these conditions: about 0·5 per cent of the total carbon put into the experiment turned up as HCN, almost the largest single product of the electron bombardment. Adenine, (HCN)$_5$, is also an important product, together with urea and lactic acid. The main point here is that HCN, adenine, and imidazole (5-aminoimidazole carboxamide, a product on the way to adenine) resulted.

(c) *Thermal experiments*

The last of the four energy inputs we have discussed is thermal energy. The experiments for this have not been as complete or as detailed as they have been for electrical discharge, but they have been somewhat more detailed than for the ultraviolet light. The experiment was carried out by passing the gas mixture through a hot tube, with a fairly long contact time (many seconds), and collecting the products in a solution. This experiment has been done by several people; the one I am reporting here was done by Sidney Fox.[12] Fig. 6.5 shows Fox's thermal apparatus, which

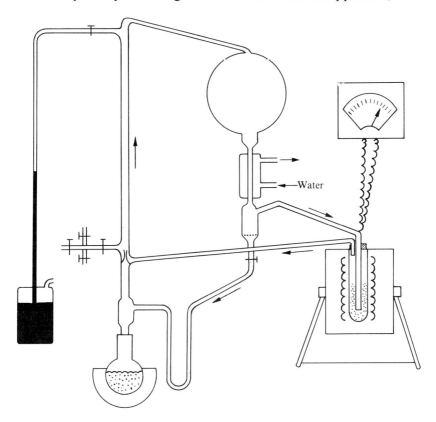

FIG. 6.5. Closed apparatus for cycling of reaction gases.[12]

is almost of the same type as that used by Miller for the electrical discharge work, except that the reaction chamber, instead of being a sparking chamber, is a hot tube. Again there is a circulating system, with the gases going through the hot tube and being pulled back through the U-tube into the boiling ammonia. The methane is thus pumped round. The hot furnace may have in it a variety of packings; one of them was silica, another was alumina. The products are collected in ammonia and heated. Fox analysed primarily for amino acids, and Table 6.5 gives the results for some of his experiments with thermal synthesis, comparing them with those obtained by electrical discharge (Miller's experiments). The products are dominated by glycine and alanine at 950°, but their concentration falls somewhat at 1050°, with glutamic acid and serine becoming dominant.

TABLE 6.5

Compositions of amino acids produced thermally in the presence of silica and by electric discharge†

Amino acid	Thermal synthesis			Electric discharge synthesis	
	Quartz sand (950 °C) (%)	Silica gel (950 °C) (%)	Silica gel (1050 °C) (%)	Spark discharge‡ (%)	Silent discharge‡ (%)
Aspartic acid	3·4	2·5	15·2	0·3	0·1
Threonine	0·9	0·6	3·0	—	—
Serine	2·0	1·9	10·0	—	—
Glutamic acid	4·8	3·1	10·2	0·5	0·3
Proline	2·3	1·5	2·3	—	—
Glycine	60·3	68·8	24·4	50·8	41·4
Alanine	18·0	16·9	20·2	27·4	4·7
Valine	2·3	1·2	2·1	—	—
Alloisoleucine	0·3	0·3	1·4	—	—
Isoleucine	1·1	0·7	2·5	—	—'
Leucine	2·4	1·5	4·6	—	—
Tyrosine	0·8	0·4	2·0	—	—
Phenylalanine	0·8	0·6	2·2	—	—
α-NH₂ butyric acid	0·6	—	—	4·0	0·6
β-Alanine	?§	?§	?§	12·1	2·3
Sarcosine	—	—	—	4·0	44·6
N-Methylalanine	—	—	—	0·8	6·5

† Basic amino acids are not listed in the table, because these amino acids have not been fully studied. Some analyses of the thermal products showed peaks corresponding to lysine (ornithine) and arginine.

‡ Recalculated from the results of Miller (1955).

§ β-Alanine peak obscured next to another unknown peak.

More quantitative measurements on experiments of this type are beginning to show differences between the various types of energy inputs, as well as indicating the mechanisms involved.

Hydrogen cyanide polymerization

I should at this point like to describe again a juxtaposition of events that began about 1960 and led to a new stage in our work, and in the work of others as well. At about that time a combination of a variety of facts and observations focused our attention on HCN as a more important and versatile molecule, beyond its function in a Strecker synthesis as described above. What were these facts that focused our attention on HCN? First of all, HCN appears extremely easily in the electrical discharge experiments, and it does not matter much what is put in. In other words, the methane–ammonia–hydrogen–water system and the carbon monoxide–nitrogen–hydrogen system (studied first by Abelson in about 1960) both produced a great deal of HCN.[13] Abelson found a large amount of HCN even though he did not have methane present in his experiment. Long before that, in fact, as a result of spectroscopic evidence we had known, in comet tails, of the presence of the C≡N molecule, which is one of the more complex molecules seen in comet spectra.[14] There were some who thought that most of the organic matter on the surface of the earth was the residue of the various comet tails trapped by the earth in its earlier history: I think, perhaps, this is a little extreme and not really a necessary presumption. Third, the appearance of HCN in the electron-bombardment experiments in water also emphasized its importance. In 1960 the discovery by Oró of the presence of adenine, $(HCN)_5$, in the water-soluble residue from the polymerization of concentrated HCN and NH_3 focused attention once more on the importance of HCN.[15–17]

At about that time we were beginning to think in our own laboratory of the next stage of chemical evolution, namely, the need to combine the various amino acids, sugars, purines, lipids, etc. into polymeric structures. It occurred to me that the carbon–nitrogen triple bond, with three pairs of electrons holding the two atoms together, might be sufficiently specifically unsaturated to be used to extract water from certain components in the same way that carbodiimide, a reagent well known to organic chemists, does. Carbodiimide contains multiple carbon–nitrogen bonds, —N=C=N—, and is useful in extracting water from monomeric materials and hooking the monomers together.[18] Recognizing the correspondence between (or the similarity of) the multiple bonds between the carbon and nitrogen in HCN and those in the carbodiimides, and recognizing also that the parent compound of the carbodiimide is cyanamide in a tautomeric form, we thought that HCN could function without a secondary transformation and remove water by adding the elements of water across the two π bonds to the nitrogen to give the same type of coupling reaction as does carbodiimide, accompanied by the hydrolysis of HCN to formamide. This would not be a simple hydrolytic reaction but a multiple-stage reaction in which the oxygen and hydrogen come, not from water, but from the corresponding monomers that are to be joined together. We had already focused our attention on HCN for these several reasons, and its importance was finally re-emphasized by the appearance of certain amino acids in the same reaction mixture in which adenine was formed from the interaction of HCN with NH_4OH. This result was reported shortly after by several people who studied the solution that remained after the black precipitate from this NH_4CN was filtered.[19] The very amino acids we have spoken of before—glycine alanine, aspartic acid, serine—occurred in this mixture, as a result of the interaction of HCN with concentrated ammonia at elevated temperatures.

In connection with the regular appearance of HCN and the evidence of its having been formed in most of the primeval atmosphere experiments, either from methane–ammonia–hydrogen–water, or CO–nitrogen–hydrogen, it is interesting to note that the free energy of the reaction

$$CH_4 + NH_3 \rightarrow HCN + 3H_2$$

drops as the temperature rises. It turns out that at room temperature, about 300 °K, the free energy of this reaction is about $+44$ kcal. As the reaction temperature rises, the energy required to make the reaction go decreases, and at 1000 °K the free energy is only $+3$ kcal. If we extrapolate a little, we find that at 1050 °K the free energy is 0; and if the temperature is raised a little higher HCN becomes the more stable compound in the system. If in addition there is a removal of hydrogen by gravitational escapes it is apparent that the reaction will proceed well, and at lower temperatures.

The fact that HCN will actually be thermodynamically stable at temperatures above 1000 °K is also, I think, of importance because of the various ways in which HCN might have made its appearance on the primitive earth. HCN will also have a negative free energy of formation from H_2, N_2, and CO if the water that is formed in that reaction is frozen out in a low-temperature layer below the upper atmosphere in which the reaction could be accomplished by the high temperatures achieved by radiation and electrical discharge.

$$H_2 + N_2 + CO \rightarrow HCN + H_2O\downarrow$$
$$\text{Frozen out}$$

The appearance of many of the nitrogen-containing compounds in the earlier primitive atmosphere experiments is now emerging as being due to the presence of HCN. There are the amino acids,[20] the purines and pyrimidine bases,[15–17,21–27] and possibly even the porphyrins,[28] an extremely important group of compounds. The structure of the porphyrin molecule, which is necessary for all sorts of energy manipulations by living cells, whether they be photochemical or oxidative, is as follows:

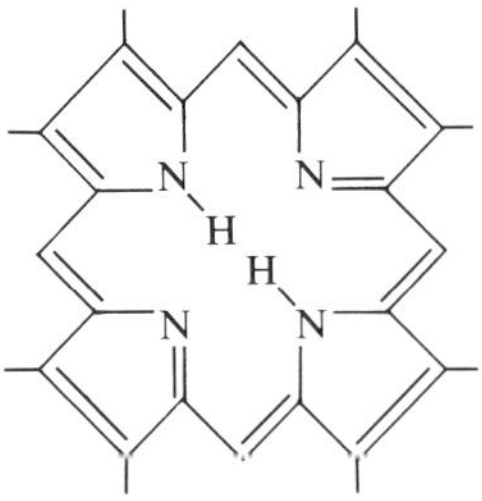

The porphyrins are the very important pigments that ultimately give rise to chlorophyll and haem. Recently, Hodgson has reported the identification of porphyrins from the methane–ammonia–water–hydrogen electric discharge; the amounts of porphyrins resulting from this are small, and the mechanism is as yet unknown, but the porphyrins are nevertheless there, and have been identified by chromatography, by their absorption spectra and fluorescence spectra, and by metal complexing.[28]

What are the possible ways in which the HCN can produce this variety of materials? In going back through the literature one finds that HCN is a rather reactive material, not only in the Strecker reaction, but with itself as well. The general polymerization of HCN by itself in water has been known for at least 150 years, but what specifically is formed in that reaction has not. Only in the last ten years or so has any really significant work been done on the nature of this polymer of HCN. It is usually a dark, or black, material,[19] which comes out of the solution after a suitable period of polymerization, almost always catalysed by some kind of base, either ammonia itself, hydroxide ion, or a variety of other Lewis bases. Recently not only has the study of the HCN polymerization reaction taken on a new interest, but it has also been demonstrated that it can be photo-induced as well as thermally induced. In Chapter 5 we discussed a thermal shock experiment (using the shock tube), and in that experiment I would predict (from the foregoing calculations) that a single pass would give about a 50 per cent yield of HCN if methane–ammonia alone had been the gases. Assume that this has happened, and that we have collected this HCN in water, with a little extra ammonia present as a catalyst, and rained down out of the upper atmosphere, thus giving a dilute solution of NH_4CN (in the ocean), with sunshine coming down on the solution. We can simulate such a system in a quartz vessel containing sodium cyanide and ammonia at a pH of about 9. A quartz-mercury lamp can simulate the sun. This is a forced run—chemical evolution in a hurry! As the reaction warms up, with the light on, the experiment will be a combined thermal and photo-chemical experiment.

We can now discuss the initial phases of the HCN reaction. Two polymers of HCN have been well characterized: the trimer and the tetramer.[29,30] The characterization of the dimer is presumably under way; a reaction scheme for this polymerization is shown opposite.[31] The reaction can be formulated in a variety of ways, including a simple polymerization of a metastable carbene, cyano-amino-carbene, a tautomeric form of the as yet unknown dimer.

The trimer that results from this reaction is amino malononitrile, and from it another carbon–nitrogen double bond can be created, which can add another HCN to get the established structure for the tetramer of HCN, which is also known. From there on, the reactions are unknown, and various suggestions have been made for the structure of the dimer. The creation of an indefinite series of carbon–carbon bonds as a result of the polymerization of HCN has long been in the literature. If that were all there was to it, the adenine and the amino acids would have to come by an independent route; in fact, the suggestion is that amino acid (glycine) comes from the amino malononitrile by a complete hydrolysis of one of the nitriles, to give the carboxyl group, which could decarboxylate and give amino acetonitrile. This could either hydrolyse to glycine (which probably would not occur) or polymerize to polyglycinimide, which is likely. This material could be partly hydrolysed, presumably, to give a polyglycine derivative that would, upon complete hydrolysis, produce glycine. This is the basic idea suggested by Akabori; that amino acetonitrile could be the raw material for polyglycine and that polyglycine was the first real peptide, or protein, in chemical evolution.[32] What Akabori wanted to do after that was to use the 'active' hydrogen of polyglycine, and insert other groups (in general via a carbonyl function) to build up side chains on the polyglycine. This suggested sequence of reactions beginning with trimer (aminomalononitrile) is shown on p. 138.

The polymers of HCN as they have so far been described vary in size, up to molecular weights of about 1000. I think that methods for handling the larger (higher molecular weight) polymers are now evolving and that in the future we shall know materials of higher molecular weight.

The formation of the adenine, which appeared before the porphyrin, seems to come through the HCN trimer, the steps for which are shown in Fig. 6.6. Another HCN is added to the amino group of the trimer to form the imidazole, containing the five-membered ring (amino-cyano-

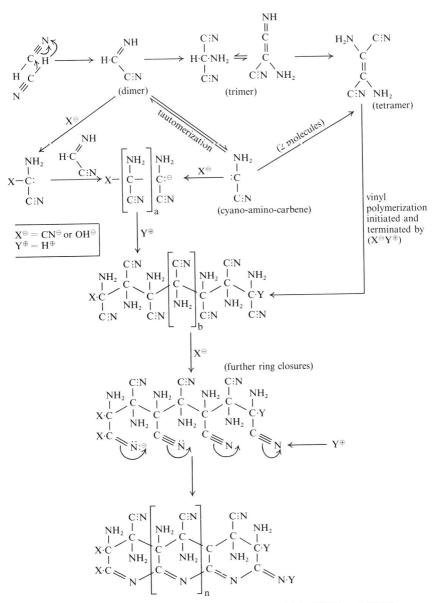

n, the number of tetramers, seems to vary between 2 and 6, i.e.(HCN)$_8$ and (HCN)$_{24}$

$$
\underset{\text{amino malononitrate}}{\overset{\displaystyle C\equiv N}{\underset{|}{H_2N-CH-C\equiv N}}}
\xrightarrow{+H_2O}
\overset{\displaystyle CO_2H}{\underset{|}{H_2N-CH-C\equiv N}}
\xrightarrow{-CO_2}
\underset{\text{amino acetonitrile}}{H_2N-CH_2-C\equiv N}
$$

polymerization ↘   ↓ +H₂O

$$
\underset{\text{polyglycinimide}}{H_2N-CH_2-\overset{NH}{\overset{\|}{C}}-\left[NH-CH_2-\overset{NH}{\overset{\|}{C}}\right]_x NH-CH_2-C\equiv N}
\qquad
\underset{\text{glycine}}{H_2N-CH_2-CO_2H}
$$

↓ +H₂O   ↑ +H₂O

$$
\underset{\text{polyglycine}}{H_2N-CH_2-\overset{O}{\overset{\|}{C}}-\left[NH-CH_2-\overset{O}{\overset{\|}{C}}\right]_x NH-CH_2-\overset{O}{\overset{\|}{C}}-OH}
$$

↓ +RCH=O

$$
-\left[NH-\underset{\underset{\displaystyle R}{\overset{|}{\underset{|}{H-C-OH}}}}{CH}-\overset{O}{\overset{\|}{C}}- \right] \quad \text{various polypeptides}
$$

imidazole), which upon hydrolysis (shown in Fig. 6.7) gives a structure such that reaction with formamide, obtained by the hydrolysis of HCN, gives adenine. The last step in the condensation, to give the adenine, is yet to be experimentally established.[29]

In 1963 Markham at Cambridge recognized that a large proportion of the amino acids formed in the simple thermal polymerization reaction were actually present in some combined form in the polymer itself.[20] The earlier analyses of Oró, Akabori, and others were not done in such a way as to recognize that the amino acids that were present were originally not present as free amino acids. Markham recognized this, as he did one more thing. He placed in such a reaction mixture some synthetic ^{14}C-labelled amino acids and was able to show that the ^{14}C-labelled amino acids turned up in the polymeric materials that he could isolate from the soluble components. In fact, he obtained three radioactive polymers from the reaction mixture and was able to show that they varied in their amino-acid composition, shown in Table 6.6. These molecules were

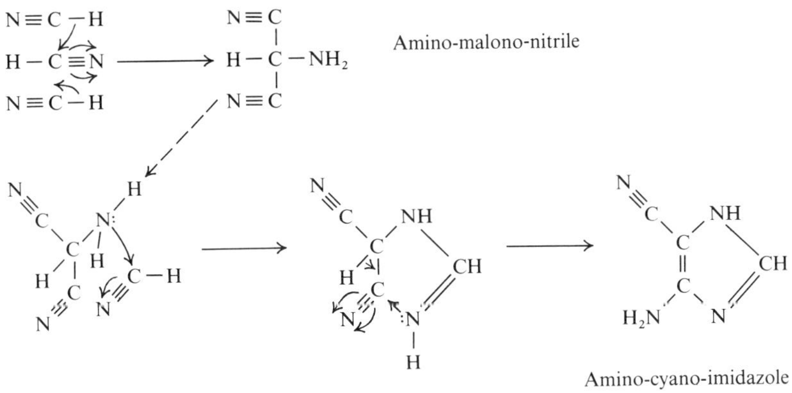

FIG. 6.6. Formation of adenine via HCN trimer.

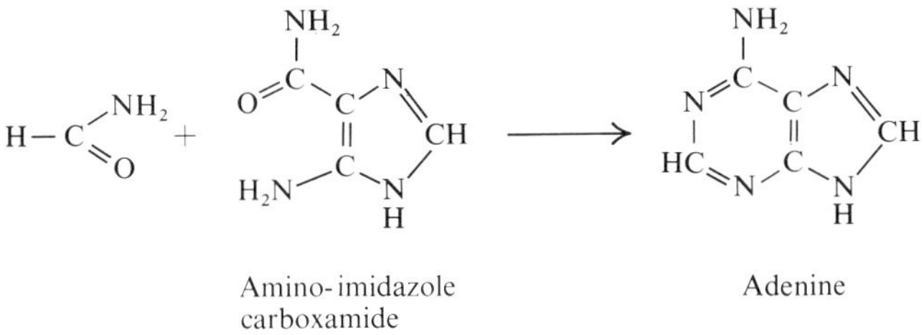

FIG. 6.7. Synthesis of adenine under primitive earth conditions.

made by putting in some labelled glycine (a trace); the massive amount of glycine obtained was a result of the polymerization. Glycine, again, dominates in the mixture.

TABLE 6.6

Radioactivity-containing polymers isolated from the reaction-mixture of polymerizing HCN in the presence of a trace of ^{14}C glycine

(Molar ratios)			
	Polymer 1	Polymer 2	Polymer 3
Glycine	67	47	134
NH$_3$	243	420	135
Asp	2·5	1·3	
Ala	13	2	

In the third polymer it appears that there is one mole of ammonia for every mole of glycine. This is surprising, because it is hard to imagine polyglycine, or any kind of glycine peptide, that would have that kind of structure. This required some explanation, which is not as yet at hand unless it be that the polyglycinimide is sufficiently stable to be isolated. Markham thus recognized the fact that there might very well be a dehydrating mechanism for actually pulling the preformed amino acids into some kind of polymeric constitution.

More recently, a new suggestion has been made for the course of the HCN polymerization reaction, at least for the thermal reaction.[33] It is suggested that the dimer, which was used previously, might actually be functioning in quite a different manner as a carbene.

The carbene structure has a variety of electromeric arrangements, including one in which a pair of electrons can be moved to form a carbonium ion structure, thus creating a 1,3-dipolar ion. Also, a pair of electrons could be moved from the nitrogen to the carbon atom, creating another electromeric form, the 1,4-dipolar ion. It has been suggested by Matthews that the 1,3-dipolar ion, a particular resonance form of the carbene, can form the CN polymers, leading eventually to polyglycine, following reduction. The reduction would probably go via hydrolysis of a cyanide and decarboxylation. Note that the stoichiometry of cyanide hydrolysis, followed by decarboxylation, is the equivalent of reduction by formic acid, with its terminal oxidation to carbon dioxide. This type of reaction can be repeated over and over again on the structures, so that the side chains of the amino acids can be built up. Some of these relationships are shown in the reaction scheme opposite.

In our laboratory demonstration of the polymerization of HCN, our accelerated chemical evolution, the first visible result is the appearance of a yellow colour in the solution, followed by the formation of a black precipitate. The nature of this black material requires investigation. Undoubtedly a mixture of all these polymeric systems is contained in the material and the various kinds of resulting molecules will depend on the temperature, the pH, the concentration, the amount of light, and perhaps other variables as well. All these things will affect the degree to which each one of these various kinds of processes operates.

It is quite clear that a wide variety of substances are formed in this type of HCN polymerization: at least 50 different materials have so far been separated by Markham, and 10 or 15 have been identified; and this does not include the macroscopic polymer type.

The mechanism of this process is certainly not established; there may be more than one, in fact. But it is already clear that HCN and ammonia constitute a very versatile source of many of the expected materials upon which the next stages of chemical evolution can operate. These next stages will be the hooking together of many of the small molecules

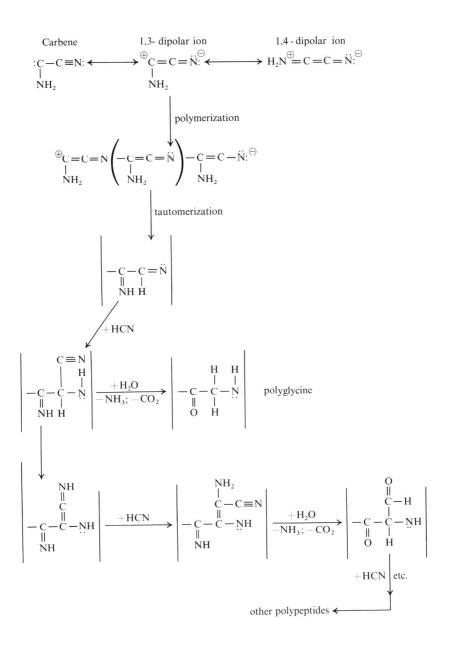

generated in this stage. We shall now go on to discuss the assembly of the entire collection of monomeric molecules into macromolecular structures.

REFERENCES

1. UREY, H. C., *The planets*. University of Chicago Press (1952).
2. OPARIN, A. I., *Origin of life*, English edition, p. 101. Dover, New York (1953).

Two recent reviews of abiogenic synthesis are references 3 and 4:

3. PONNAMPERUMA, C. A., and GABEL, N. W., Current status of chemical studies on the origin of life. *Space Life Sci.* **1**, 64 (1968).
4. LEMMON, RICHARD M., Abiogenic synthesis of biologically relevant organic compounds (chemical evolution), *University of California Lawrence Radiation Laboratory Report*, UCRL 18108 (1968); *Chem. Rev.*, in press.
5. GARRISON, W. M., MORRISON, D. C., HAMILTON, J. G., BENSON, A. A., and CALVIN, M., Reduction of carbon dioxide in aqueous solutions by ionizing radiation. *Science* **114**, 416 (1951).
6. MILLER, S. L., Production of amino acids under possible primitive earth conditions. Ibid. **117**, 528 (1953).
7. GROTH, W. E., and WEYSENHOFF, H. v., Photochemical formation of organic compounds from mixtures of simple gases. *Planet. Space Sci.* **2**, 79 (1960).
8. TERENIN, A. N., Photosynthesis in the shortest ultraviolet. *Origin of life on the earth* (editor A. I. Oparin), pp. 136–9. Pergamon Press, London (1959).
9. MILLER, S. L., Production of some organic compounds under possible primitive earth conditions. *J. Am. chem. Soc.* **77**, 2351 (1955).
10. —— The mechanism of synthesis of amino acids by electric discharge. *Biochim. biophys. Acta* **23**, 488 (1957).
11. PALM, C., and CALVIN, M., Primordial organic chemistry. I. Compounds resulting from irradiation of $^{14}CH_4$. *J. Am. chem. Soc.* **84**, 2115 (1962).
12. HARADA, K., and FOX, S. W., The thermal synthesis of amino acids from a hypothetically primitive terrestrial atmosphere. *The origins of prebiological systems and of their molecular matrices* (editor S. W. Fox), p. 190. Academic Press, New York (1964).
13. ABELSON, P. H., Chemical events on the primitive earth. *Proc. natn. Acad. Sci. U.S.A.* **55**, 1365 (1966).
14. MILLER, S. L., and UREY, H. C., Organic compound synthesis on the primitive earth. *Science* **130**, 245 (1959).
15. ORÓ, J., Synthesis of adenine from hydrogen cyanide. *Biochem. biophys. Res. Commun.* **2**, 407 (1960).
16. —— and KIMBALL, A. P., Synthesis of purines under possible primitive earth conditions. I. Adenine from HCN. *Archs Biochem. Biophys.* **94**, 217 (1961).
17. —— —— Synthesis of purines under possible primitive earth conditions. II. Purine intermediates from HCN. Ibid. **96**, 293 (1962).
18. KHORANA, H. G., *Some recent developments in the chemistry of phosphate esters of biological interest*. Wiley, New York (1961); specific references here to p. 33.
19. WADSTEN, T., and ANDERSEN, S., Studies on a polymerization product of hydrogen cyanide. *Acta chem. scand.* **13**, 1069 (1959); LABADIE, M., JENSEN, R., and NEUZIL, E. Recherches sur l'évolution pré-biologique. III. Les acides azulmiques noirs formes à partir du cyanure d'ammonium. *Biochim. biophys. Acta*, **165**, 525 (1968).

20. Lowe, C. U., Rees, M. W., and Markham, R., Synthesis of complex organic compounds from simple precursors: formation of amino acids, amino acid polymers, fatty acids and purines from NH₄CN. *Nature, Lond.* **199,** 219 (1963).

21. Ponnamperuma, C. A., Mariner, R., Lemmon, R. M., and Calvin, M., Formation of adenine by electron irradiation of methane, ammonia and water. *Proc. natn. Acad. Sci. U.S.A.* **49,** 735 (1963).

22. —— —— and Sagan, C., Formation of adenosine by ultraviolet irradiation of a solution of adenine and ribose. *Nature, Lond.* **198,** 1199 (1963).

23. —— and Kirk, P., Synthesis of deoxyadenosine under simulated primitive earth conditions. Ibid. **203,** 400 (1964).

24. —— and Mack, R., Nucleotide synthesis under possible primitive earth conditions. *Science* **147,** 1221 (1965).

25. Oró, J., Synthesis of organic compounds by electric discharge. *Nature, Lond,* **197,** 862 (1963).

26. —— Studies in experimental organic cosmochemistry. *Ann. N.Y. Acad. Sci.* **108,** 464 (1963).

27. Morita, K., Ochiai, M., and Marumoto, R., A convenient one-step synthesis of adenine. *Chemy Ind.* August 1968, p. 1117.

28. Hodgson, G. W., and Ponnamperuma, C. A., Prebiological porphyrin synthesis: porphyrins from electric discharge in methane, ammonia and water vapor. *Proc. natn. Acad. Sci. U.S.A.* **59,** 22 (1968).

29. Sanchez, R. A., Ferris, J. P., and Orgel, L. E., Studies in prebiotic synthesis. II. Synthesis of purine precursors and amino acids from aqueous HCN. *J. molec. Biol.* **30,** 223 (1967).

30. Ferris, J. P., Sanchez, R. A., and Orgel, L. E., Studies in prebiotic synthesis. III. Synthesis of pyrimidines from cyanoacetylene and cyanate. Ibid. **33,** 693 (1968).

31. Matthews, C. N., and Moser, R. E., Prebiological protein synthesis. *Proc. natn. Acad. Sci. U.S.A.* **56,** 1068 (1966).

32. Akabori, S., Origin of the fore-protein. *The origin of life on the earth* (editor A. I. Oparin), p. 189. Pergamon Press, London (1959).

33. Matthews, C. N., and Moser, R. E., Peptide synthesis from hydrogen cyanide and water. *Nature, Lond.* **215,** 1230 (1967).

7

SELECTION AND THE GROWTH
OF MOLECULES

WE have seen a rather extensive sequence of reactions that showed how it was possible to derive, by a variety of energy inputs, a whole set of the small molecules upon which living organisms of today depend. In principle, this random synthesis could not go on indefinitely and get very far, because the same energy inputs that dismantle the primeval molecules and allow them to rearrange themselves into more interesting precursor molecules would eventually, as these biological precursors accumulated, destroy them. Some additional selection process must be introduced, apart from the separations mentioned earlier. The separation processes were physical separation of the products from the energy input (for example, by rain, which brings the products down from the upper atmosphere and thus into a region of less ultraviolet flux) or an analogous process that would move them out of the electrical-discharge region. These physical separation processes would in themselves not be enough to provide for the survival of very complex material. In addition, it is clear that some kind of molecular selection process must obtain that can be brought about by *autocatalysis*, a term and process that has been known in chemistry for many years.

Autocatalysis

This process of autocatalysis (and eventually stereospecific autocatalysis) plays a major role in molecular selection.[1] Autocatalysis is indeed the chemical term for a self-reproducing system, a property that the biologists are prone to use as one of the criteria for a living system. The primitive catalytic properties of many of the simplest molecules can be ramified, eventually producing the highly specific and efficient catalytic systems we now know in biological processes and are accustomed to associate with enzymes.

We shall now examine some of the primitive catalytic properties of both ions and simple molecules and determine how they could in practice, or at least conceptually, in principle, be elaborated into highly specific reactions. Fig. 7.1 is a representation of a sequence of catalytic materials that was selected for several specific reasons. Hydrogen peroxide is a very common product of the impact of ultraviolet light or ionizing radiation on water. The question is whether or not it is possible to devise effective useful decomposition of this peroxide that might be autocatalytic in its function. As Fig. 7.1 shows, ferric ion has a very meagre catalytic coefficient, viz. 10^{-5}. This figure represents the efficiency of the ferric ion in decomposing hydrogen peroxide. If that ferric ion is incorporated into a porphyrin molecule, which is basically a tetrapyrrole ring with various side chains on it, the efficiency of the resulting iron porphyrin molecule in decomposing hydrogen peroxide goes up by a factor of 1000, from 10^{-5} to 10^{-2}. If, in addition to this, the iron porphyrin molecule is incorporated still further into a polypeptide chain, particularly the protein component of catalase, a specific protein that places an imidazole ring on either side of the iron, then the catalytic efficiency goes up by further factor of 10 000.[1,2] The efficiency of this essential catalyst, the iron ion, which can accept and donate an electron (according to its condition), has thus been enhanced by something like eight or nine powers of 10 by these several changes. I am not suggesting that these changes occurred in just these three steps, since it is conceivable that there are many intermediate steps in which the catalytic power of the iron is enhanced by some smaller factor.

The question that now arises is: what kind of a system could perpetuate this—how could this fact be used in a molecular selection process? The appearance of the porphyrin ring, the macrocyclic tetrapyrrole ring, has been described recently as coming from the simple molecules of the primitive atmosphere in several different ways. It was described some years ago as forming spontaneously from δ-aminolaevulinic acid upon irradiation. It was described more recently as arising from more primitive materials, starting with the simple molecules methane–ammonia–water–hydrogen.[3] After this atmosphere has been subjected to electric discharge, various yields of porphyrins have been reported. Also, it is known that the yield of the porphyrin in this very simple primitive system can be enhanced by the addition of a divalent metal ion. Some 10 to 15 years ago, when we first began the study of

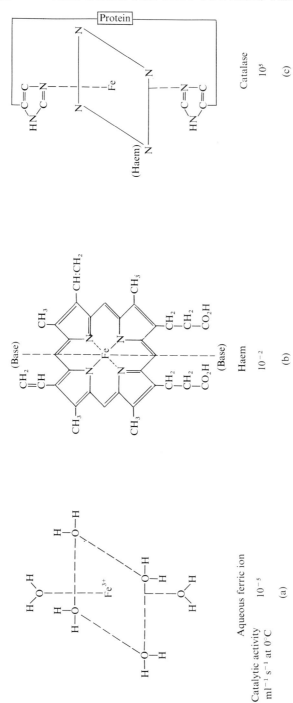

Fig. 7.1. Evolution of a catalyst for the reaction $H_2O_2 \rightarrow H_2O + \frac{1}{2}O_2$.

porphyrin synthesis, we had used the addition of divalent metal ions to enhance the yields of porphyrins formed from the condensation of an aldehyde and a pyrrole; a corresponding phenomenon was recently described again by those who did the electric-discharge experiments. Given in Fig. 7.2 is a scheme of the construction of a porphyrin from simple compounds; formaldehyde, an aldehyde, that appears in all the electric discharge experiments, and pyrrole, which can be generated in a similar way. There are four pyrrole rings and four aldehyde-bridging

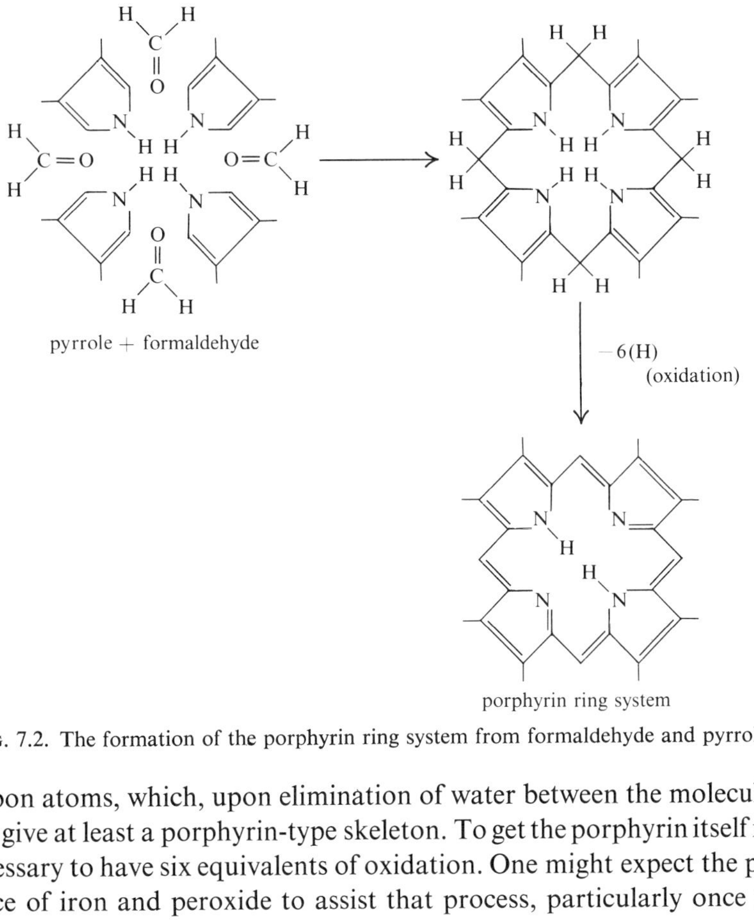

pyrrole + formaldehyde

−6(H)
(oxidation)

porphyrin ring system

FIG. 7.2. The formation of the porphyrin ring system from formaldehyde and pyrrole.

carbon atoms, which, upon elimination of water between the molecules, will give at least a porphyrin-type skeleton. To get the porphyrin itself it is necessary to have six equivalents of oxidation. One might expect the presence of iron and peroxide to assist that process, particularly once the process starts and the iron gets into the haem, which is the resultant of that oxidation step, and which in turn makes a better oxidant—a better catalyst for the oxidation—of the initial condensation product. This mechanism would select for the generation of porphyrins out of such a system. This is simply an example of an autocatalytic system that we

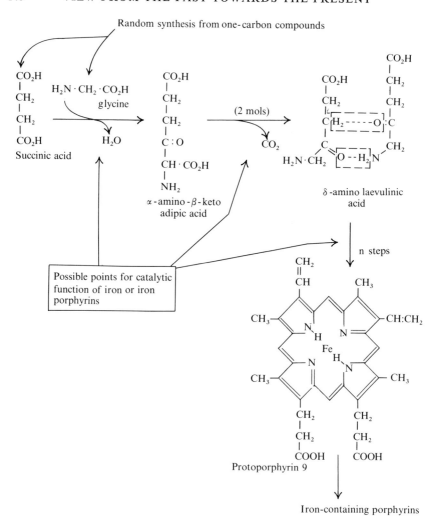

Fig. 7.3. The present-day biosynthetic pathway of the porphyrins. Protoporphyrin 9 is the isomer of protoporphyrin usually found in biological systems.

can put together in the laboratory and whose function we can demonstrate experimentally. I suspect that the same kind of function can be found in the present-day route for biosynthesis of porphyrins, which is shown in Fig. 7.3.[4] δ-aminolaevulinic acid can, as mentioned above, be condensed by radiation, in the presence of suitable divalent metal cations, to produce an appreciable yield of protoporphyrin. Several steps are involved in this condensation, and oxidation is also involved; the functioning of the iron and iron porphyrins as catalysts in these oxidations is indeed demonstrable. Here, again, there is the same kind of auto-catalytic function that would give rise to a selective generation of those particular structures. One can devise for other systems similar auto-catalytic processes. This mechanism is sometimes called a *reflexive catalysis* rather than autocatalysis, because the final product does not necessarily have to function as a catalyst in the last direct reaction that gives rise to it. It can function as a catalyst in a reaction further back, giving rise to one of its precursors. Perhaps the term *reflexive catalysis* is a more accurate description of the process, since it is not merely a simple autocatalytic reaction.

Regularly in discussions of this kind the question of the stereo-specificity of biological systems arises. Usually what is thought of in this terminology is the optical specificity that present-day biological systems show. Optical specificity relates to the fact that almost all biological materials have in them the possibility of being right- or left-handed; at the molecular level the reactions existing in the biological system generally work on only *one* of these two mirror-images.[5] The fact that biological systems do indeed exhibit this degree of optical specificity has been relied upon as evidence for the participation of a living organ-ism in the generation of anything that is asymmetric. Today, in general, the generation of an optically pure material in the laboratory requires the intervention of a living organism, either a microbe or a man. Some-where in the sequence there is a living organism. This basic idea is generally introduced in all first-year organic chemistry teaching.

What I am going to discuss has required of me a certain amount of 'readjustment'; I am now going to say that it is possible to generate optically pure material without the intervention of a microbe or a man, provided that certain conditions are fulfilled. In fact, what I say is that optical activity in the products of such a chemical evolutionary system as we have been developing cannot help but arise the moment we intro-duce the notion of *stereospecific autocatalysis*. The moment we require (or allow) this kind of a process to participate in the evolutionary scheme, then the racemic system (that is, the equimolar mixture of right-handed and left-handed materials) is an unstable system and must go one way or the other, depending on chance, to a first approximation. Fig. 7.4 gives a schematic diagram of what is involved. Suppose that we have a system of molecules (A) that can exist in a right- and left-handed form, and let us suppose that the right- and left-handed forms are relatively freely interconvertible, so that the rate at which they can change one into the other is quite fast. Let us also suppose that in addition to this the substance (A) can be converted into the substance (B) by some reaction r. Let $k(d)$ be the rate at which the d form of (A) is converted into the d form of (B), and $k(l)$ the identical rate at which the l form of (A) is converted into the l form of (B). Let these uncatalysed rates of conversion be very slow compared to the equilibration and compared to the time available. One other condition is imposed, namely, that the l molecule of (B) is catalytic for the reaction of conversion of l(A) to l(B); i.e. $k_{\mathrm{cat}(d)}$ and $k_{\mathrm{cat}(l)}$ will be many powers of 10 greater than $k(l)$ or

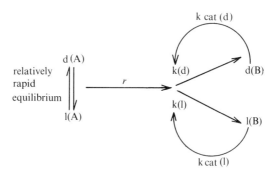

FIG. 7.4. A system with potential optical specificity.

$k(d)$. We can see that since the initial uncatalysed rates are slow, in the initial phase of the reaction the numbers of these two (B) molecules, (l) and (d), cannot be equal at the beginning. The *first* molecule that goes will be either (l) or (d); i.e. statistically at the beginning of the reaction there will be more of one than the other. If the rate constants k and $k_{\mathrm{(cat)}}$ are sufficiently different, then immediately the whole pool of (A) will transform into one or the other of these two optically active forms of (B), but not both. Thus, in a very slowly starting system there will always be a statistical dominance (at first) of one form or the other. Thus, when the product (B) is a stereospecific catalyst for its own formation, the entire product will be dominantly of the initially formed configuration.

One of the earliest examples of this type of reaction that I encountered is shown in Fig. 7.5.[6] The tetrahedral asymmetric structure given in Fig. 7.5 for the quaternary amine is potentially optically active. In the reaction we have equimolar amounts of the d and l forms in solution; there is a rapid reversible dissociation into the tertiary $R_1 R_2 R_3$-amines and allyl X. The quaternary amine is optically active and the tertiary one is not; there is thus a means for interchange between d and l forms. If the solvent conditions in which the reaction is taking place are adjusted so that the material begins to crystallize out slowly, the d and l forms do not crystallize together. The first crystal nucleus that forms, if it happens to be a d crystal, will cause all the material to come out as d crystals. The same phenomenon occurs if the first crystal is an l crystal: all the material will come out as l crystals. This type of reaction illustrates the basic requirement in all its simplicity. The nucleated solution containing the crystal of d (or of l) is the catalyst for its own crystallization, and the trick is to adjust the solvent conditions so that crystallization does not occur too rapidly. Repeated independent crystallizations produce pure forms of either d or l crystals, with about equal probability.

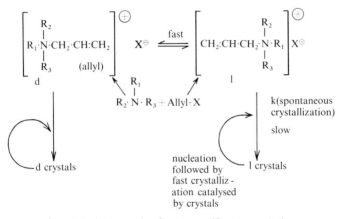

FIG. 7.5. An example of stereospecific autocatalysis.

This particular experiment is not particularly interesting in biological terms except to illustrate a principle. However, the potential for a very important biological autocatalytic case has come to light by the work of Gillard and Allen, which has to do with the use of octahedral cobalt complexes that seem to fulfil all the required conditions.[7] Details of this reaction are shown in Fig. 7.6. The asymmetric molecule $[(Co(en)_2(H_2O)(OH)]^{2+}$ has all the qualities that are required for optical activity. But it has some other remarkable properties as well. If this molecule is exposed to a dipeptide, an interesting reaction occurs. It turns out that if the initial reactive molecule in the equilibrium is either d or l, the product being unable to equilibrate between the d and l forms, only one stereoisomer is formed. The reaction is stereospecific on the cobalt. Here we see a possibility for an autocatalytic system as well, if the ligands of the products are now capable of being displaced by another peptide system (the second peptide). The d form will be 'frozen in' and it will select from the racemic peptides only one, and not the other. Thus we have an autocatalytic system for selecting one of the optical forms of the peptide, to give a stereospecific degradation of that peptide. I am introducing this idea here because it will also be relevant to a later discussion. It has all the potential qualities that are required for spontaneous stereospecific evolution of the type we imagine to be required. This hydrolytic reaction can be reversed. If, instead of starting with a peptide and an aquohydrate to give an amino acid, we start with an amino acid ester and a cobalt dichlorotetramine, we can build up a

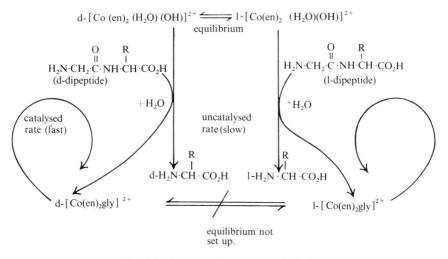

FIG. 7.6. Stereospecific peptide hydrolysis.[7]

dipeptide, according to the actual conditions and amounts of the activated amino acids present.[8] This reaction can thus be used either for degradation or construction.

   In general, then, not only can we see the primitive catalytic qualities of simple ions and molecules evolving, as shown in the specific examples discussed above, but we can see that the optical specificity is an intrinsic property of the kinds of systems (atoms, molecules, and reactions) with which we are dealing.

Growth of molecules

We have traced the basic mechanisms for making all the simple molecules that are necessary for a living system, and we have discussed the idea of autocatalysis and stereospecific autocatalysis for selecting among the simple molecules formed by these random, or statistical, processes.

Before we can perform any selective degradation of, say, the peptides, as mentioned above, we have to create the peptide. The specific example of the latter process just brought forward uses an 'activated' amino acid. We have not so far demonstrated the general ways of allowing these simple molecules, which are created initially by the various energy inputs to the primeval materials, to grow into more complex molecules. The

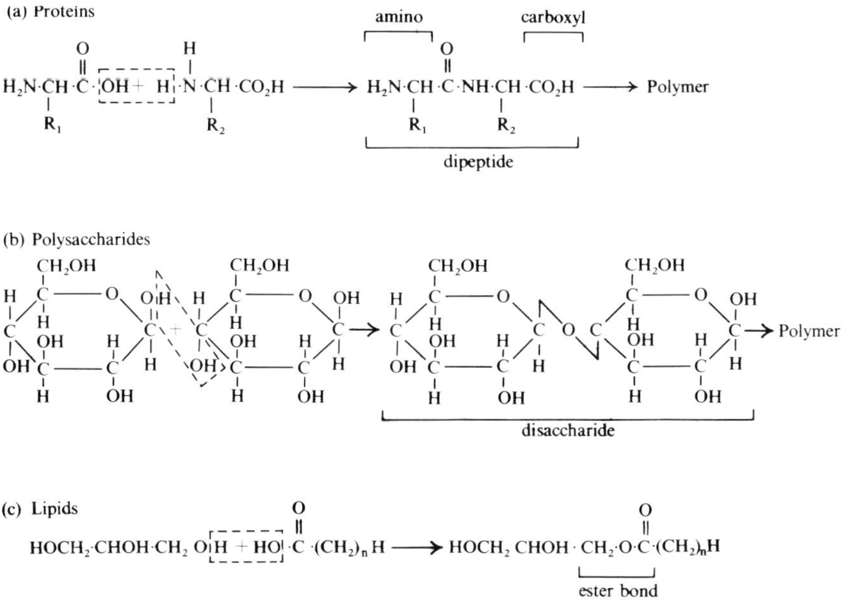

FIG. 7.7. Dehydration condensation reactions in the formation of proteins, carbohydrates, and lipids.

growth of all the types of biopolymers in which we are interested (proteins, nucleic acids, polysaccharides, lipids) involves a dehydration condensation. I shall next describe this type of reaction for each of the major classes of biopolymers: proteins, nucleic acids, polysaccharides, and lipids.

Fig. 7.7 gives a generalized description of the dehydration condensation reactions. Proteins are constructed by a dehydration reaction between the carboxyl end of one amino acid (the acid end) and the amino end (the basic end) of another. Removal of the elements of water between these two ends, resulting in a bond formed between two molecules that are originally separate, produces a dipeptide. We are left with the same two functional groups, i.e. a carboxyl group and an amino group, at the ends of the molecule, and these can grow provided that some system is devised for inducing the dehydration condensation. For the formation of polysaccharides, the problem is identical (see Fig. 7.7 (b)). There is a hydroxyl group on the semiacetal carbon (carbon atom 1) of a simple sugar, and there are various other hydroxyl groups with which it could condense to form a link to create the disaccharide; this gives more variety than does the amino acid condensation because of

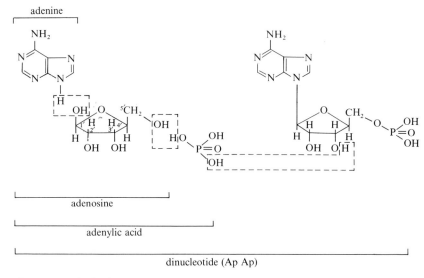

FIG. 7.8. Dehydration condensation reactions in the formation of ribonucleic acid. DNA lacks the OH group on the 2′ carbon of the ribose.

the greater numbers of hydroxyl groups that can be used to give different branches. For the lipids, exactly the same kind of problem exists—the formation of a carbon–oxygen bond by elimination of a hydroxyl group from the carboxylic acid and a hydrogen atom from an alcoholic hydroxyl group to form a carboxylic ester. With the lipids the product molecule does not have two of the original functional groups; this is why the lipids, in general, are not usually included in the class of macromolecules. Their formation does, however, involve exactly the same kind of dehydration condensation reaction that is used to form the proteins and polysaccharides. Because of their apparent widespread importance in the formation of cellular structures, we have included them here.

   The dehydration condensation of the nucleic acids is shown in Fig. 7.8. Here again, it is apparent that dehydration condensations can occur in three different places. First, to form the nucleoside from the heterocyclic base (adenine, which we created so easily from hydrogen cyanide) and the hydroxyl group of a ribose sugar. This forms the riboside, or desoxy-riboside, linkage to the nitrogen. The next dehydration condensation involves the formation of a phosphate ester between the primary alcohol group of the terminal carbon atom of the ribose and one of the hydroxyl groups of phosphoric acid. We now have a nucleotide, a three-component monomer: the base, the sugar, and the phosphoric acid. In effect, only two of these components are involved in the succeeding reaction; one of the remaining hydroxyl groups of the phosphoric acid and one of the remaining hydroxyl groups of the sugar are used to link one nucleotide with another to give a dinucleotide. We end up exactly as before with a phosphoric-acid function at one end of the molecule and a hydroxyl function at the other, so that the molecule can grow at either end just as the protein can.

There is one thing common to all these reactions: they involve the sticking together of some of these small molecules, all of which have been generated by random synthesis, by the same kind of process—the removal of water between two parts of the same kinds of molecules.[9]

Various schemes have been devised for achieving these dehydration condensation reactions to create, for example, polypeptides, nucleic acids, and other macromolecules. Fig. 7.9 (a) shows in somewhat greater detail what the polypeptide primary structure of a protein can be. The backbone of the polypeptide is identical to one of those suggested as resulting from the photopolymerization of hydrogen cyanide. The photo- and thermal-polymerization of hydrogen cyanide gives polymers whose structure has not yet been completely identified, and it may be that it is possible to achieve such a protein-like backbone. It remains to be seen whether this is the best method of achieving it.

Fig. 7.9 (b) also shows some of the many possibilities for amino-acid composition.

Thermal dehydration condensation reactions

One of the condensation methods that was very early suggested as playing a possible role in the evolution of proteins or polypeptides was the rather obvious one of heating up a mixture of these amino acids, each of which has the amino and carboxyl function, to a temperature at which the water that would result from a condensation between them would simply distil out of the reaction-mixture, thus creating the pep-tides. The trouble with this type of experiment is that most of these amino acids do not melt very easily, and it is difficult to get them to react when they are dry and crystalline. The resulting material is usually a carbonized tar. It was not until Sidney Fox in Florida noted that he could make the amino acids melt by taking a mixture and essentially using one or two of the amino acids as a solvent that peptides could be easily created in this way.[10-12] The pair of amino acids that Fox used successfully for this purpose were aspartic and glutamic acids. He found

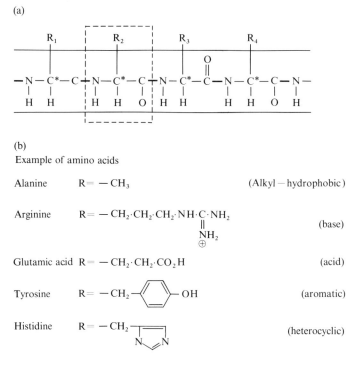

(a)

(b)
Example of amino acids

| Alanine | R= —CH₃ | (Alkyl—hydrophobic) |

Alanine $R = -CH_3$ (Alkyl — hydrophobic)

Arginine $R = -CH_2 \cdot CH_2 \cdot CH_2 \cdot NH \cdot C \cdot NH_2$ (base)

Glutamic acid $R = -CH_2 \cdot CH_2 \cdot CO_2 H$ (acid)

Tyrosine $R = -CH_2$—⟨ ⟩—OH   (aromatic)

Histidine $R = -CH_2$ (heterocyclic)

FIG. 7.9. (a) The primary structure of a protein and (b) examples of the main classes of amino acids.

that if he took a mixture of these two (roughly 50 per cent), then added the other 20 amino acids (roughly 50 per cent), and heated the mixture up to about 150 °C he got a good melt, and the water would distil out. Then, if he plunged the resulting reaction-mixture into water, he could obtain what he called 'proteinoid' from it. Fox went one step further. Recognizing that in present-day biological systems a particular kind of linkage (in effect the phosphoric anhydride linkage, which is a particularly good dehydrating agent) is used to remove the water, Fox was able to make the reaction go at temperatures below 100 °C by using the reaction-mixture of amino acids and carrying out the reaction in a polyphosphoric acid solution. This was made from aqueous phosphoric acid by heating at a given temperature to remove a certain amount of water from it. After a given time a variety of polyphosphates was formed, and this resulting material was used as a solvent for the amino-acid reaction at temperatures below 100 °C.

Table 7.1 shows the analysis of the crude proteinoid mixture that Fox obtained from such a reaction.[13] In this particular reaction the starting

TABLE 7.1

Amino acid composition of proteinoid prepared with 200 ppa†[13]

	(%)		(%)
Thr	0·55	Lys	2·79
Ser	0·63	His	2·53
Pro	1·04	Arg	1·83
Gly	2·93	Total	7·15
Ala	1·31		
Val	1·33	Asp	51·9
Met	0·86	Glu	13·3
Iso	0·71		
Leu	3·44	Total	65·2
Tyr	3·87		
Phe	5·87		
NH_2	5·02		
Total	27·6		

† Asp:Glu:basic and neutral amino-acid ratio = 2:1:3; 100 °C, for 150 h.

material was 33 per cent aspartic acid, 17 per cent glutamic acid, and the remaining 55 per cent consisted of some 17 amino acids, about 3 per cent each. Table 7.1 shows that the insoluble proteinoid does not have the composition of the starting mixture of amino acids. Instead of 33 per cent aspartic acid, the proteinoid contains 51 per cent; instead of 17 per cent glutamic acid, 13 per cent; and instead of 3 per cent of each of the other amino acids, the composition varies a great deal. This proteinoid formation is not a statistical reaction; there is some selectivity. There must be selection principles built into this mixture of amino acids to create the results obtained by Fox, which is the major point that Fox makes. He has also taken the crude reaction-product and fractionated it by gel filtration. This is a system of fractionating the dissolved material by passing it through a column containing insoluble materials in a gel form whose holes are of the right specific size; as the solution goes through this column packed full of gel, the small molecules enter the holes in the gel but the larger ones cannot and therefore come through the column first. This gives a pattern of proteinoid coming out at the bottom of the column with many peaks, not a uniform broad distribution as one might expect if the distribution were purely statistical. Fox has compared his column eluate of proteinoid with a column eluate of serum protein, and they have a generally similar appearance, with similar patterns.[14,15]

The point to be made here is that this combination of amino acids, even as made by this rough thermal treatment, is not a random, statistical combination. Although Fox used dehydrated phosphoric acid, more sophisticated polyphosphates have been used in similar types of reaction, as described in the following section.

Use of polyphosphate esters in dehydration condensation reactions

The first more sophisticated polyphosphate that was used in this type of reaction was made by dissolving phosphoric anhydride (P_4O_{10}) in diethyl ether, to form an ethyl polyphosphate.[16] The synthesis of this ethyl polyphosphate has been known for a long time, but its use as a condensing agent was first achieved with significant results in 1962 by Schramm. The formation and possible structure of 'polyphosphate ester' is shown in Fig. 7.10. On the left is the structure of the phosphoric anhydride (P_4O_{10}), which is really a symmetrical tetrahedral structure. In diethyl ether the phosphoric anhydride structure opens up at one of the bridging oxygens to form the diethyl cyclic polyphosphate.[17,18] The reaction then goes on to form a variety of other materials. These materials are now useful for performing the dehydration condensation reaction, using the P—O—P link to take up water, to form essentially two molecules of phosphoric acid. The use of this soluble phosphoric anhydride linkage has now been extended to other systems.[19] But the reaction must be carried out, at least to a first approximation, in an anhydrous solvent, one that contains no water. The usual solvent used is ethyl ether or dimethylformamide. This same reaction can be done with a slightly more stable polyphosphate, using a phenyl group instead of two methyl groups, to yield essentially the same kind of product, in which the polyphosphates are linked to phenyl groups. The phenyl phosphates can be used in aqueous solution, although the yields are quite low.

Schramm has taken adenine and ribose in dimethylformamide solution, at roughly a few millimolar concentration, and has obtained a 20 per cent yield of adenosine; he has taken adenine and deoxyribose and obtained a 40 per cent yield of both alpha and beta deoxyadenosine. In the first case he used the ribose sugar, in which every carbon atom has a hydroxyl group on it, and in the second he used deoxyribose, in which the carbon 2 does not have a hydroxyl group. Schramm also carried out the reaction with the phenylpolyphosphates in water; the yields were only about 3 per cent, but, nevertheless, he did obtain a product in water.

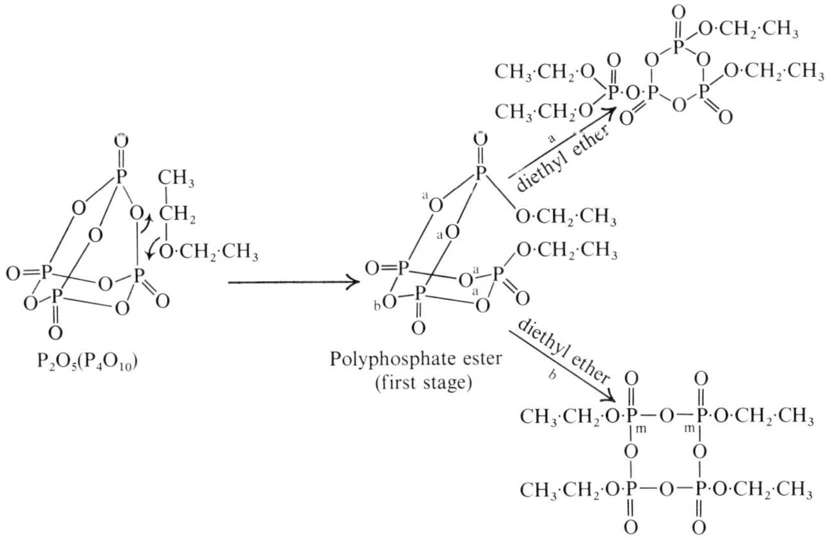

FIG. 7.10. The formation and possible structure of 'polyphosphate ester'.

Starting with the nucleoside, or nucleotide, Schramm was able to obtain polynucleotides, i.e. he achieved the hooking together of the nucleotides to form nucleic acids in the manner shown in Fig. 7.8. The analysis of the polynucleotides obtained by this method, using enzymes specific for the 5' phosphate or the 3' phosphate, showed that both were present.

These reactions show a need to have anhydrous or polymeric media, and high-temperature conditions, and they would require certain rather special geological conditions on the primitive earth. For example, we must presume evaporation to dryness in periodically wet places, such as tidal pools, in order to obtain the kind of chemical environment in which this kind of reaction would go. The intervention of volcanic heat would probably also be necessary to perform these reactions even in polyphosphoric acid. These conditions are special enough to warrant a search for dehydrating condensation reactions in aqueous solution in a more general way, more or less as they occur at present. I might add that in all living systems all of these dehydration condensation systems are going on and in a water milieu.

The next stage in our work at Berkeley was thus to begin the search for chemical conditions that could produce this dehydrating condensation reaction in water itself. In other words, we wanted to take out water between two molecules, but do it in an aqueous environment! This is chemically a tricky thing to do; it requires an energy input, because the reaction is thermodynamically 'uphill'. In present-day living organisms the energy input is supplied by ATP (adenosine triphosphate), with its phosphoric anhydride linkage. This type of phosphoric anhydride linkage is used in modern biology to achieve this and many other thermodynamically uphill results, and the energy thus supplied is spoken of as the *energy source* for the organism.

We had to search for another mechanism for bringing about these dehydration condensation reactions that might have preceded pyrophosphate linkage in the evolutionary scheme as the primary condensing agent. I want now to refer to a fact that I mentioned earlier: there is another way in which energy is stored as a result of these primary inputs. This is in the form of the carbon–nitrogen multiple bond, as hydrogen cyanide or its derivatives. The first of these derivatives is cyanamide ($H_2N \cdot C:N$), which will dimerize (one amino group adding to one cyanide) to give the dicyandiamide, and, finally, another compound of the same family, the dicyanamide (two cyanides on one nitrogen atom). All these compounds have carbon–nitrogen bonds and all are capable of absorbing a water molecule to form a $C{=}O$ and an $N{-}H$ group and a variety of other products as well.

We shall next discuss the kinds of materials that can be generated using the carbon–nitrogen multiple bond as it appears in cyanamide.[9] This is the tautomer of the parent compound that the organic chemists call carbodiimide, in which, in general, the hydrogens have been replaced by groups of carbon atoms. We shall also discuss the various achievements of this carbodiimide-type molecule, which is readily generated by any of the primeval energy inputs, and show that it can make pyrophosphates, peptides, phosphate esters, and carboxy esters. The mechanisms of these syntheses will be examined. With the low reaction-temperature in a mild condensation reaction of this sort, selectivity can be seen even more readily. There is a possibility of a non-template specific protein formation in this way.

REFERENCES

1. CALVIN, M., Chemical evolution and the origin of life. *Am. Scient.* **44**, 248 (1956).
2. —— Evolution of the photosynthetic apparatus, *Science* **130**, 1170 (1959); Evolution of photosynthetic mechanisms, *Perspect. Biol. Med.* **5**, 147 (1962).
3. HODGSON, G. W., and PONNAMPERUMA, C. A., Prebiological porphyrin synthesis: porphyrins from electric discharge in methane, ammonia and water vapor. *Proc. natn. Acad. Sci. U.S.A.* **59**, 22 (1968).
4. SHEMIN, D., Biosynthesis of porphyrins. *Harvey Lect.* **50**, 258 (1954).
5. WALD, G., The origin of optical activity, *Ann. N.Y. Acad. Sci.* **69**, 352 (1967); GARAY, A. S., Origin and role of optical isomery in life, *Nature, Lond.* **219**, 338 (1968).
6. HAVINGA, E., Spontaneous formation of optically active substances. *Biochim. biophys. Acta* **13**, 171 (1954).
7. ALLEN, D. E., and GILLARD, R. D., Stereoselective effects in peptide complexes. *Chemical communications* 1967, p. 1091.
8. COLLMAN, J. P., and KIMURA, E., The formation of peptide bonds in coordination sphere of cobalt (III). *J. Am. chem. Soc.* **89**, 6096 (1967).
9. For a more detailed discussion of dehydration condensation reactions, see STEINMAN, G., Ph.D. thesis, University of California, Berkeley, December 1965; Protobiochemistry: dehydration condensation in aqueous solution, *University of California Lawrence Radiation Laboratory Report* UCRL 16566 (1967).
10. FOX, S. W., and HARADA, K., Thermal copolymerization of amino acids to a product resembling protein. *Science* **128**, 1214 (1958).
11. —— —— and VEGOTSKY, A., Thermal polymerization of amino acids and a theory of biochemical origin. *Experientia* **15**, 81 (1959).
12. —— —— Thermal copolymerization of amino acids common to protein. *J. Am. chem. Soc.* **82**, 3745 (1960).
13. —— and YUYAMA, S., Abiotic production of primitive protein and formed microparticles. *Ann. N.Y. Acad. Sci.* **108**, 487 (1963).
14. —— Prebiological formation of biochemical substances. *Organic geochemistry* (editor I. A. Breger), pp. 36–49. Macmillan, New York (1963).
15. —— Simulated natural experiments in spontaneous organization of morphological units from proteinoid. *Origins of prebiological systems and of their molecular matrices* (editor S. W. Fox), pp. 361–73. Academic Press, New York (1964).
16. SCHRAMM, G., GROTSCH, H., and POLLMANN, W., Nonenzymatic synthesis of polysaccharides, nucleosides, nucleic acids and the origin of self-reproducing systems. *Angew. Chem.* int. Edn. **1**, 1 (1962).
17. BURKHARDT, G., KLEIN, M. P., and CALVIN, M., The structure of the so-called 'ethylmetaphosphate' (Langheld ester). *J. Am. chem. Soc.* **81**, 591 (1965).
18. SCHRAMM, G. and POLLMANN, W. Reactivity of metaphosphate esters prepared from P_4O_{10} and ethyl ether. *Biochim. biophys. Acta* **80**, 1 (1964).
19. —— LÜNZMANN, G., and BECHMANN, F., Synthesis of α and β-adenosine and of α and β-deoxyadenosine with phenyl polyphosphate. Ibid. **145**, 221 (1967).

8

GROWTH OF MOLECULES AND
INFORMATION COUPLING

Aqueous dehydration condensation reactions

WE have been examining ways in which the monomers, the simple small molecules that we could generate abiologically, could be hooked together to form the macromolecules that we know are the basis of today's living organisms. Among others, we made use of a variety of compounds related to hydrogen cyanide, one of the principal materials that appears in all the primary energy input experiments.

Earlier, I also illustrated the way in which the multiple carbon–nitrogen bond in effect stored the primary energy input to be used later in the dehydration condensation reaction, which is essentially the type of re-action with which we have to deal.[1] These potential storage compounds include a number of cyanide derivatives. We have done most of our experiments with cyanamide and dicyandiamide. Cyanamide, H_2N—CN, formed from HCN and NH_3, is a tautomer of H—N=C=N—H (carbodiimide); dicyandiamide is formed from cyanamide.

$$HCN + NH_3 \rightarrow H_2N \cdot C \vdots N \rightleftharpoons HN \vdots C \vdots NH$$
$$\text{cyanamide} \qquad \text{carbodiimide}$$

$$\underset{\substack{\| \\ NH}}{H_2N \cdot C \cdot NH \cdot C \vdots N} \rightleftharpoons \underset{\substack{\| \\ NH}}{H_2N \cdot C \cdot N \vdots C \vdots NH}$$
$$\text{dicyandiamide}$$

The third derivative of the high-energy input reactions is a molecule in which two of the hydrogens of ammonia have been replaced by cyanide groups to form dicyanamide, which is also capable of exhibiting this same carbodiimide tautomerism:

$$N \vdots C \cdot NH \cdot C \vdots N \rightleftharpoons N \vdots C \cdot N \vdots C \vdots NH$$

The —N=C=N— is common to all three of these structures. This peculiar arrangement makes it possible for the molecule to grasp, on the central carbon atom, an oxygen atom from whichever base may be available. The molecule will also react with a nitrogenous base. It reacts particularly well with phosphoric acid or carboxylic acid. The equations below summarize these reactions; probable mechanisms are shown later.

$$(1) \quad \underset{\underset{\text{H}_2\text{N}\cdot\text{CH}\cdot\text{C}\cdot\text{OH}}{}}{\overset{\overset{\text{R}_1 \quad \text{O}}{|\qquad\|}}{}} + \underset{\underset{\text{H}_2\text{N}\cdot\text{CH}\cdot\text{CO}_2\text{H}}{}}{\overset{\overset{\text{R}_2}{|}}{}}$$

$$\xrightarrow[\quad]{\cdot\text{N}:\text{C}:\text{N}\cdot} \underset{\underset{\text{H}_2\text{N}\cdot\text{CH}\cdot\text{C}\cdot\text{NH}\cdot\text{CH}\cdot\text{CO}_2\text{H}}{}}{\overset{\overset{\text{R}_1 \quad \text{O} \qquad\quad \text{R}_2}{|\qquad\| \qquad\qquad |}}{}} + \underset{}{\overset{\overset{\text{O}}{\|}}{\cdot\text{NH}\cdot\text{C}\cdot\text{NH}\cdot}}$$

$$(2) \quad \overset{\overset{\text{O}}{\|}}{\text{A}\cdot\text{O}\cdot\underset{\underset{\text{OH}}{|}}{\text{P}}\cdot\text{OH}} + \overset{\overset{\text{O}}{\|}}{\text{HO}\cdot\underset{\underset{\text{OH}}{|}}{\text{P}}\cdot\text{O}\cdot\text{B}} \xrightarrow[\quad]{\cdot\text{N}:\text{C}:\text{N}\cdot} \overset{\overset{\text{O} \quad\; \text{O}}{\| \qquad \|}}{\text{A}\cdot\text{O}\cdot\underset{\underset{\text{OH OH}}{|\quad\;\; |}}{\text{P}\cdot\text{O}\cdot\text{P}}\cdot\text{O}\cdot\text{B}} + \overset{\overset{\text{O}}{\|}}{\cdot\text{NH}\cdot\text{C}\cdot\text{NH}\cdot}$$

$$(3) \quad \text{R}\cdot\text{OH} + \overset{\overset{\text{O}}{\|}}{\text{HO}\cdot\underset{\underset{\text{OH}}{|}}{\text{P}}\cdot\text{O}\cdot\text{B}} \xrightarrow[\quad]{\cdot\text{N}:\text{C}:\text{N}\cdot} \overset{\overset{\text{O}}{\|}}{\text{R}\cdot\text{O}\cdot\underset{\underset{\text{OH}}{|}}{\text{P}}\cdot\text{O}\cdot\text{B}} + \overset{\overset{\text{O}}{\|}}{\cdot\text{NH}\cdot\text{C}\cdot\text{NH}\cdot}$$

$$(4) \quad \text{R}\cdot\text{OH} + \overset{\overset{\text{O}}{\|}}{\text{HO}\cdot\text{C}\cdot\text{R}_1} \xrightarrow[\quad]{\cdot\text{N}:\text{C}:\text{N}\cdot} \overset{\overset{\text{O}}{\|}}{\text{R}\cdot\text{O}\cdot\text{C}\cdot\text{R}_1} + \overset{\overset{\text{O}}{\|}}{\cdot\text{NH}\cdot\text{C}\cdot\text{NH}\cdot}$$

Thus, in the case of peptide formation, (1) above, once the central carbon atom of the carbodiimide form has been connected with an oxygen atom of the carboxyl group, the new carbon–oxygen bond is very strong with the unshared electron pair of the oxygen atom being drawn into the remaining C=N system. The reaction then proceeds by further nucleophilic attack upon the carboxyl carbon atom, assisted by the additional tendency of the π-pair of electrons of the remaining carbonyl group to migrate into the electronegative carbonyl oxygen, leaving the carbon atom as an incipient carbonium ion subject to nucleophylic attack by the amino group of another amino acid.

dipeptide   urea

We have thus formed a peptide link, and the urea that comes out of this reaction is a very stable material. There are also many side reactions to be dealt with, but I shall not discuss here the ramifications of this chemistry. I want to point out the versatility of the CN group and its relatives including the cyanate ion, and how we can get these reagents from the primitive atmosphere. Not only can we introduce a dehydration condensation between an amino group and a carboxyl group, but, as the reaction sequences above show, we can do it with a whole variety of potentially polymer-forming materials, such as the formation of pyrophosphate itself, the formation of a phosphate ester, and even the coupling of carboxylate with alcohol to form a carboxylate ester, the fatty-acid esters of the lipids. All these have been demonstrated by the use of one or other of the reagents dicyanamide, cyanamide,[2,3] dicyandiamide,[4] and cyanate.[5]

One of the most important of these results is the creation of pyrophosphate linkages from orthophosphate. Orthophosphate in a suitably ionized form can undergo a dehydration coupling to give pyrophosphate:

$$
\underset{\overset{|}{OH}}{HO-\overset{\overset{O}{\|}}{P}-OH} + \underset{\overset{|}{OH}}{HO-\overset{\overset{O}{\|}}{P}-OH} \rightarrow \underset{\overset{|}{OH}}{HO-\overset{\overset{O}{\|}}{P}-O-}\underset{\overset{|}{OH}}{\overset{\overset{O}{\|}}{P}-OH} + H_2O
$$

In effect we have thus created the essential molecular grouping that is now the principal agency by which the dehydration condensations (including polymerizations) are mediated in present-day living organisms. Here, then, we have a way of creating pyrophosphate through the intermediacy of carbon–nitrogen multiple bonds. There are alternative ways of obtaining pyrophosphate linkages from primitive types of materials, and a number of experimenters have devised and tested other methods. Miller tested potassium cyanate on the surface of the mineral calcium hydroxyapatite to obtain a good yield of pyrophosphate; ammonium and calcium phosphates have also been used, with excellent results.[5]

$$
KNCO + \underset{\text{(hydroxyapatite)}}{Ca_{10}(PO_4)_6(OH)_2} \overset{\Delta}{\longrightarrow} \text{pyrophosphate}\downarrow
$$

$$
+ (NH_4)_3PO_4 \longrightarrow \text{pyrophosphate}\downarrow
$$

$$
+ Ca_3(PO_4)_2 \longrightarrow \text{pyrophosphate}\downarrow
$$

Another possible route to pyrophosphate is the photochemical one, sensitized by the heterocyclic bases: pyridine, phosphoric acid, and calcium when illuminated by ultraviolet light should give pyrophosphate in reasonable yield. This is a reaction in which the ultraviolet energy is directly utilized, via sensitization by the heterocyclic base. Other heterocyclic compounds will do the same thing.[6]

$$\text{(pyridine)} + H_3PO_4 + Ca^{2+} \xrightarrow[\text{ultraviolet light}]{h\nu} \text{pyrophosphate}$$

Finally, the oxidation of ferrous iron by hydrogen peroxide, which is one of the products of the radiation decomposition (activation) of water, also produces a small yield of pyrophosphate.[7]

$$Fe^{2+} + H_3PO_4 + H_2O_2 \rightarrow Fe^{3+} + \text{pyrophosphate}$$

I wish only to mention that there are other ways, apart from those involving cyanide derivatives, of generating the pyrophosphate or phosphoric anhydride linkage that today, as we know, is the primary route by which living organisms accomplish dehydration condensations.

Fig. 8.1 shows a possible mechanism for the formation of peptides with hydrogen cyanide as a dehydrating agent; this was the first of this

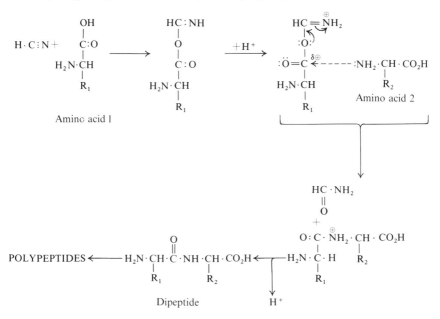

FIG. 8.1. A mechanism of peptide formation with hydrogen cyanide as the dehydrating agent.

group of CN compounds to be suggested. It does not matter whether dicyanamide, cyanamide, or dicyandiamide is used as the dehydration condensation reagent; all have an actual or potential carbodiimide linkage, and the differences are in the co-products of the dehydration condensation. In the case of dicyandiamide, the product actually is a guanylurea. Fig. 8.2 shows a presumed mechanism for peptide formation using dicyandiamide as the reagent.[4] The mechanism of peptide formation using dicyanamide as the dehydrating agent is shown in Fig. 8.3; the end-product in this case is the dipeptide diglycine. The demonstration of all these dehydration condensations has been achieved

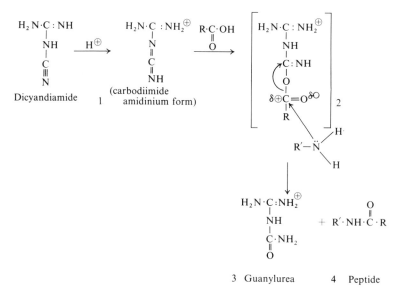

FIG. 8.2. A mechanism for peptide synthesis with dicyandiamide as the dehydrating agent.

in the laboratory, and it has gone further than a simple dipeptide.[8] Fig. 8.4 shows the polymerization of glycine, by slow addition of dicyanamide, as a function of time. We begin with free glycine, and drip into it at a constant rate a solution of the dicyanamide at slightly acid pH. The figure shows the various products that can be extracted, going as far as tetraglycine.

I wish to show that in terms of geological time this coupling reaction is adequate. It can be achieved in somewhat acidified water solution at a measurable rate; in less acid solution the rates, of course, are slower. While the resistance or restraint that we feel about the use of such reactions in evolutionary sequences may to some extent be justified, it is not restricted to the point that we must discard these hypotheses because of the long time-element involved. One of the problems of all experiments of this type, designed to reproduce what might have been evolutionary reaction-systems that took place over very long periods of time, is the slowness of some of the reactions we have to call upon. Another problem is the possible variety that may and does occur, which must be followed or accompanied by selection.

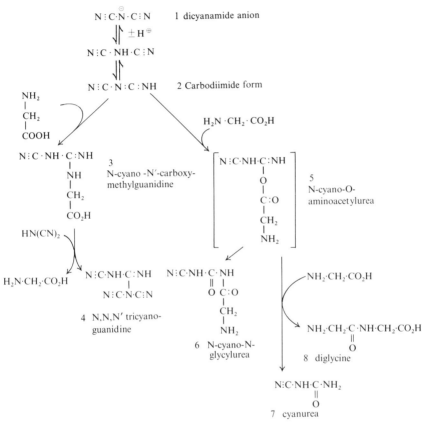

FIG. 8.3. A mechanism for peptide synthesis with carbodiimide as the dehydrating agent.

The coupling reaction that involves the coupling of two amino acids to give peptide seemed to me, when we first encountered it in aqueous media, a likely means of producing some kind of selective sequencing of amino acids. We set out to test the relative efficiency of the various amino acids in this coupling reaction with respect to their ability to come together; our aim was to see if there was any selectivity in the coupling reaction itself. We knew that there must be some selectivity, and it was only a matter of determining exactly how much—how great were the differences between the different amino acids, even though all the chemical functions are, to a first approximation, the same.

Gary Steinman designed a method of testing this selectivity in a special and simple form. He began by hooking the first amino acid on

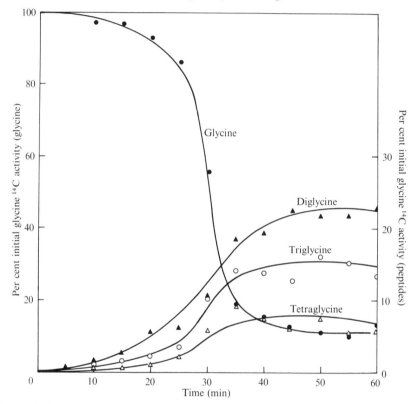

FIG. 8.4. Rates of glycine utilization and polypeptide accumulation in a solution originally containing [2-^{14}C] glycine (0·005 M) and HCl (0·1 M). Sodium dicyanamide (0·1 M in aqueous solution) was added to the reaction mixture at 0·68 mg/min for 1 hour. Samples were withdrawn periodically and analysed by paper electrophoresis.[8]

to the body of a large polymer bead. Chlormethylated polystyrene was allowed to react with a sodium salt of the amino acid, to produce the amino acid attached to the benzyl group of the resin as a benzyl ester. One can thus make a resin with one amino acid ($H_2N \cdot CHR_1 \cdot CO_2H$) on it and then examine the relative efficiency of coupling a whole series of other amino acids to it by using a series of amino-protected amino acids ($Z \cdot NH \cdot CHR_2 \cdot CO_2H$, etc.). The coupling of each one of the N-protected amino acids can be measured to give a relative efficiency of coupling in one direction, in which R_2 is the N-terminal group of the peptide. This type of experiment yielded results which are shown in Table 8.1. This is a comparison of experimentally determined dipeptide yields, obtained by the method just described, and doublet frequencies calculated from known protein sequences.[9] The coupling efficiency of glycyl-glycine was used as the norm, and the other efficiencies are measured relative to this. There is a variation in both yield and frequency of a factor of at least ten. Phenylalanine is one of the larger

TABLE 8.1

Comparison of experimentally determined dipeptide yields and frequencies calculated from known protein sequences

Dipeptide†	Values (relative to Gly–Gly)	
	Experimental	Calculated
Gly–Gly	1·0	1·0
Gly–Ala	0·8	0·7
Ala–Gly	0·8	0·6
Ala–Ala	0·7	0·6
Gly–Val	0·5	0·2
Val–Gly	0·5	0·3
Gly–Leu	0·5	0·3
Leu–Gly	0·5	0·2
Gly–Ile	0·3	0·1
Ile–Gly	0·3	0·1
Gly–Phe	0·1	0·1
Phe–Gly	0·1	0·1

† The dipeptides are listed in terms of increasing volume of the side chains of the constituent residues. Gly, glycine; Ala, alanine; Val, valine; Leu, leucine; Ile, isoleucine, Phe, phenylalanine. Example: Gly–Ala = glycylalanine.[9]

groups and does not couple very well; the coupling efficiency, for example, of phenylalanyl-phenylalanine was below the range that could be measured in this particular system. The third column in Table 8.1, 'calculated', requires explanation. Steinman looked into the *Atlas of protein structure*[10] to see how frequently, in all the protein structures that had been studied until a year ago, the particular doublet sequences occur; he then related them to the occurrence of diglycine in the particular protein.

The obvious deduction that one would like to make from such an observation is that there is some validity in the idea that sequencing in polypeptides may even today have at least some component that was originally self-determined, i.e. determined by the amino-acid sequence itself and not necessarily by the present coding system. Rather, the present coding system of amino acids is derived in some part at least from what was originally a self-determined growing-point amino-acid sequence.[9,11] I think it is worth mentioning here that a pentapeptide synthesis in bacteria has been described that does not go through a templating mechanism at all; this mechanism produces an antibiotic of which one end is uridine diphosphate N-acetyl glucosamine, with lactic acid attached to one of the hydroxyl groups of the glucose. To the hydroxyl group of the lactic acid is attached a pentapeptide, which is not templated; it is composed of two enzymes and its sequence is determined solely by the internal structure of the pentapeptide and the enzymes without the intervention of any m-RNA. The sequence of the peptide is:[12]

UDP–GNAc–Lactic–L-Ala–D-Glu–L-Lys–D-Ala–D-Ala

In other words, to the lactic acid the enzymes will attach only L-alanine; to the L-alanine only D-glutamic acid; and so on. This is *not* a templated pentapeptide synthesis; it is a pentapeptide synthesis by growing-chain determination with catalyst specificity. Thus, even in modern bacteria, we may have some 'residues' of the manner in which the first proteins or the first polypeptides might have been generated.

Sequence determination by 'growing-end control'

This leads to the general idea that some sequence determination by 'growing-end control' is possible. I think this was perhaps one of the principal ways in which the early polypeptides were generated. The polymer chemists have long since been using this concept of 'growing-end control' and have established that even stereospecific control (*d* against *l*) can be achieved in this way. For example, starting with propylene oxide, which has a potentially optically active asymmetric carbon atom, and inducing its polymerization with a zinc dialkyl, one gets either a chain of all *d* polypropylene oxide or all *l* polypropylene oxide, but not a mixture of *d* or *l* in the same chain.

$$CH_3-\overset{\overset{\displaystyle H}{|}}{C}\underset{\underset{\displaystyle O}{\diagdown\diagup}}{\quad}CH_2 \qquad \text{propylene oxide}$$

d or *l*

with Zn R$_2$ in methanol:

$$Zn\cdot O\cdot CH_2\cdot \overset{\overset{\displaystyle CH_3}{|}}{\underset{\underset{\displaystyle H}{|}}{C}}{}^{\oplus}\longleftarrow \overset{}{\underset{\underset{\displaystyle \delta\oplus CH}{\diagdown\diagup}}{O}}\longrightarrow CH_2$$
$$\underset{\displaystyle CH_3}{\qquad\qquad\qquad}$$

gives all *d* or all *l* polypropylene oxide

From another point of view, one might say that the catalyst has stereo-selective active sites and determines the stereospecificity of the polymer as well; this explanation is not quite compatible with the reaction described above.

A similar stereo-selectivity between d and l forms has been demonstrated in the polymerization of alanine-carboxyanhydride, initiated by methanol.[13,14]

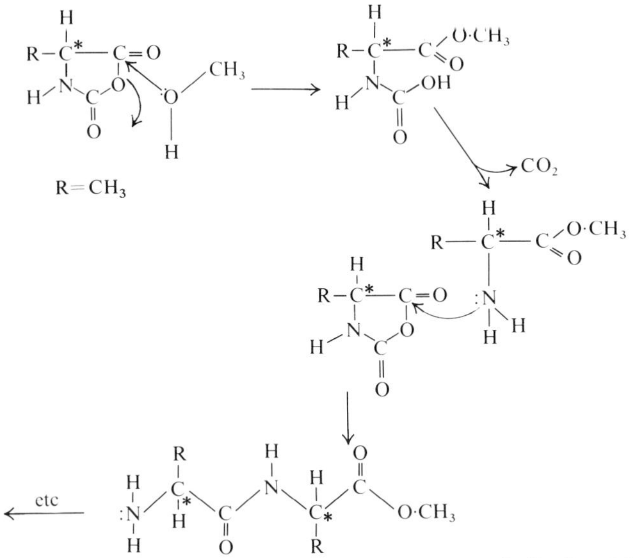

* marks the asymmetric
carbon atoms in the polymer.

Obviously one can expect a degree of selectivity among the different amino acids. The R group in the reaction shown above may be any of some 20 groups of atoms of widely differing geometry and space requirement.

Using the idea that growing-end control can indeed give rise to a relatively specific polypeptide sequence, and remembering what we learned earlier (Chapter 7) about the way in which one can obtain stereospecific and amino-acid specific peptide hydrolysis by metal complexes, much as a modern enzyme (for example, carboxypeptidase) takes off the C-terminal end of a polypeptide, one can now imagine a replicating polypeptide system in the following scheme (the capital letters represent individual amino acids):

$$\begin{array}{c} \text{H}_2\text{O} \\ \text{P}_1 \quad | \quad \text{P}_2 \\ (\text{H}_2\text{N})\,\text{A}-\text{B}\!+\!\text{C}-\text{A}-\text{D}-\text{E}-\text{M}-\text{N}-----\text{X}-\text{Y}-\text{Z} \end{array}$$

$$(\text{H}_2\text{N})\,\text{C}-\text{A}-\text{D}-\text{E}-\text{M}-\text{N}-----\text{X}-\text{Y}-\text{Z}\;+\;(\text{NH}_2)\,\text{A}-\text{B}\,(\text{COOH})\;\text{(to repeat growth)}$$

When the peptide is clipped off, by the folding back of the special catalytic end attached to the cobalt, it creates a new peptide (P_2), and the other piece (P_1) may be the beginning of another peptide chain. It is the same beginning we had before, and it can therefore grow again in a similar manner, subject of course to the statistical imprecision of the reaction and the changes in the composition of the medium. This is a polypeptide replicating system, without a template, determined by the probabilities of the various peptide linkages in the coupling reaction itself—essentially, it is growing-end control. This reaction remains to be demonstrated in detail experimentally, but I should like to refer to a recent description of what may very well be another of the 'residues' of such a system in a modern bacterium, E. coli. It has recently been noticed that most proteins in E. coli seem to start at the N-terminal end with N-formyl-methionine, followed either by alanine or serine, and then the rest of the protein.[15] When these proteins are isolated the N-formyl-methionine is usually not present. One can then imagine that that particular pair of amino acids, N-formyl-methionine with alanine or serine, might indeed have been one of the primitive starting-points of the early polypeptides that were successful in their replication mechanisms. (I introduce this only as a suggestion.) Such a mechanism as this gives us a degree of protein specificity and a protein replication system, at least in principle and, I think, eventually in experiment as well.

A system of simple monomeric linear polymers with which we are concerned is the polynucleotides. Here the mechanism of simple replication is much easier to devise and much easier to test experimentally. This is the selective sequence autocatalysis (replication system) for polynucleotides. The basic idea, of course, is the base-pairing of certain sequences of polynucleotides with the corresponding bases, together with the coupling mechanism provided by the carbodiimide or corresponding dehydrating condensing agent. The polymerization of uridine monophosphate by ethyl polyphosphate in a nonaqueous medium (usually dimethyl formamide) in the presence of polyadenylic acid provides an autocatalytic or reflexive system. Fig. 8.5 shows how the presence of polyadenylic acid acid (Curve A) enhances the rate of condensation of uridine monophosphate, as opposed to Curve B, in which the UMP is polymerized by ethyl polyphosphate in the absence of poly A. Quite clearly here the complementary base-paired polymer does indeed very greatly increase the speed of the formation of the corresponding

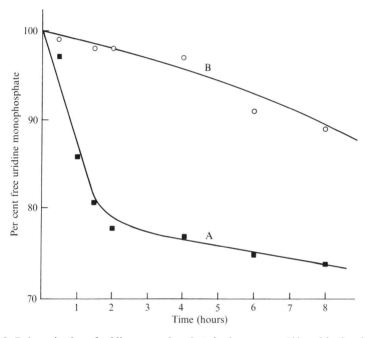

FIG. 8.5. Polymerization of uridine monophosphate in the presence (A) and in the absence (B) of polyadenylic acid. The decrease of free UMP was measured chromatographically.

polymer (poly U). This is not an aqueous system, and the likelihood of such an ethyl polyphosphate condensing system having occurred and survived in water is probably very small indeed.[16]

We therefore sought to discover condensation systems of nucleotides that would exhibit the same kind of effect in an aqueous medium. Naylor has been able to demonstrate such a reflexive catalytic effect of the corresponding bases in a system containing polyadenylic acid and polythymidylic acid of suitable chain-lengths.[17] He measured the rate of coupling of two hexathymidines to give a dodecathymidine (a 'twelve-mer'), first in the absence of any catalyst and then in the presence of a catalyst. This type of reaction is shown in Fig. 8.6. Hooking together the thymidines in water using the water-soluble carbodiimides in the absence of a catalyst gives a particular rate constant. The same experiment can be done in the presence of polyadenylic acid (poly A). (Adenine is the base that pairs with thymine.) In the presence of polyadenylic acid the thymidines 'fit' along the chain, and the probability of bringing together two hexathymidines on the template of polyadenylic acid is much greater than that of the two hexathymidines finding each other without any help. The result is that the coupling of two hexathymidines in the presence of polyadenylic acid is about 10 times as fast as it is in the absence of poly A, if the other conditions of coupling are the same. We now have a twofold system; the dodecathymidine can come free, and the same reaction can be repeated with any loose adenylic acid components that might be present, creating a catalytic system that will produce the dodecamer of adenylic acid. Upon separation of the products, we are back where we started, and thus we have a reflexive catalytic system. One of the components is not by itself autocatalytic for its own formation by itself; it has to go through one intermediate step that keeps up the replication mechanism.

We have now the two relatively independent autocatalytic systems; one for replicating a particular polypeptide or protein sequence and one for replicating a polynucleotide sequence. The polynucleotide sequence replication system has been demonstrated experimentally, but the replication of a polypeptide system such as I have proposed, even a small one of half a dozen or more linkages, has yet to be demonstrated. It must be remembered also that while the replication fidelity of a polyadenylic acid (or polynucleotide) system is high, the replication fidelity of a polypeptide will not be high. There will be a probability effect; if one allows hexapeptide to be generated in a given medium of amino acids, the same hexapeptide will not be created every time. 'Mistakes' will be made, because the probabilities of coupling are apparently not sufficiently different for the sequence to be always identically reproduced. This lack of reproducibility in the replication of a polypeptide system is important for several reasons. Although all the varieties of catalytic function are possible with protein replication, it is not sufficiently accurate; whereas with the nucleic acid replication system it is possible to have a great deal of accuracy, but the catalytic function in the system is very limited. The nucleic acid can replicate only itself, but the polypeptide, with all of its various side chains, can generate all the catalytic functional groups—acidic, basic, amphoteric, aromatic, etc.—that are necessary or useful,

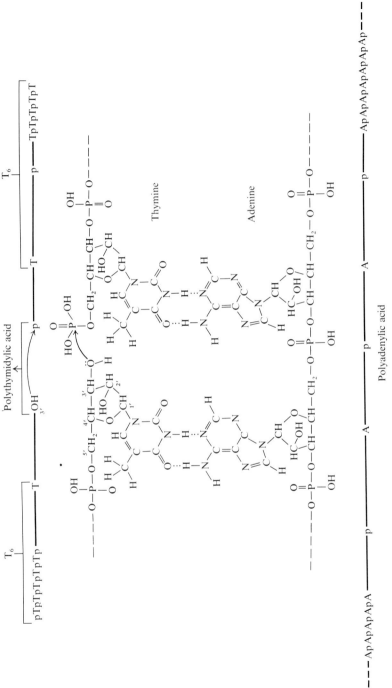

Fig. 8.6. The coupling of two hexanucleotides to form a dodecanucleotide.

because the nature of the side chains and the way they may be juxtaposed by folding imply this great potential variability.

Chemical systems—Coupling of replicating systems

We thus have two linear sequences, whose evolution presumably began independently, for different chemical reasons. If some way could be devised, or some chemical system imagined, by which these two initially independently generated linear sequences (polypeptide and polynucleotide) could be coupled so that the fidelity of the polynucleotide replication system and the multiple catalytic function of the polypeptide system could be used to best advantage, it would have great survival value.[18] It is quite clear that such a coupled reflexive system going through some multiple steps instead of just two steps, as in the polynucleotide sequence, or essentially only one step for the proposed polypeptide replicating system, would have dynamic advantages over either one alone. It might go through a sequence of different steps in which there could be three of four different stages, no one of which would be simply autocatalytic but the system itself would be a reflexive catalytic system if, for example, the last product of the series were actually catalytic for the first step. There would then be a good high-fidelity replicating system with a reasonable reaction rate. This, of course, is exactly what has happened in evolutionary terms.

Is there any chemical reason for the occurrence of such a coupling, which we can visualize as occurring spontaneously in a reaction mixture of the type we have so far considered in aqueous solution? There are other means of performing these same types of reactions, for example by adsorption on the surfaces of clay particles.[19,20] I am not going to discuss these in any detail, not because I have any *a priori* reason for believing they were improbable, but simply because I have no real experimental clues that such things happened, except perhaps the fact that kaolin clays are very absorbent and have definite patterns of charges on their layers. Other investigators have tried to use these characteristics as primitive templates.

The question of the coupling of these two linear systems—the polynucleotide system and the polypeptide system—was raised very soon after this discussion took shape. By that time we had already seen and had clearly in mind the possible peptide selectivity and also the polynucleotide replication selectivity. It was quite clear that somewhere at some time the coupling had to occur, and the question was: 'What were the chemical factors that would have produced it?' In thinking of this question, one single experimental observation of present-day organisms focused attention on a possible chemical mechanism for the generation of this coupled system. This is the fact that one of the stages in today's reflexive catalytic system, which begins with a polynucleotide sequence (DNA) and goes through a whole series of catalytic stages including feeding into another polynucleotide system (m-RNA), is the transfer of an amino acid from free solution to the terminal end of a relatively short

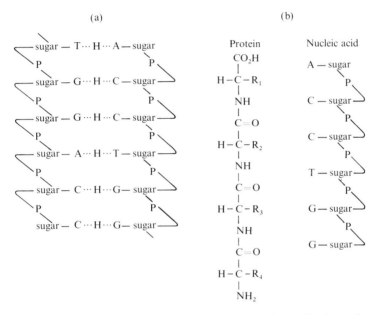

FIG. 8.7. (a) Base pairing for DNA replication and RNA template formation. (b) The linear polymeric structure of nucleic acids and proteins.

nucleic acid chain; one containing only about 100 bases. This is the transfer-RNA, a ribonucleic acid, which picks up the amino acid from the medium and eventually carries it to the place where it will be hooked into the polypeptide. Fig. 8.7 (a) shows the base pairing for DNA replication and RNA template formation, steps that precede the use of t-RNA. It is known that all the transfer RNAs have the same three bases on the amino-acid accepting end. This seemed to me a possibly significant fact, with respect to the generation of this coupled system. These three terminal bases are C—C—A (cytidine–cytidine–adenine), and it is the terminal adenine base that accepts the active amino acid on the 3′ carbon atom of its ribose residue. The active ester so formed is now subject to attack by a neighbour, constructed in the same manner, which may be on the terminal of the adjacent (t-RNA) molecule. In each case, then, the important fact is that the terminal amino acid is added to a particular triplet on the end of the t-RNA. This fact focused our attention on the possible mechanism for the generation of the coupling mechanism, and we set about testing this. If there were something peculiar about this particular base triplet end, and if exactly the same kind of selective experiments were done as we did for one amino-acid self-selection, one might find a degree of selectivity in the coupling of amino acids to bases that would depend also on the bases. Thus, we have set up an experiment

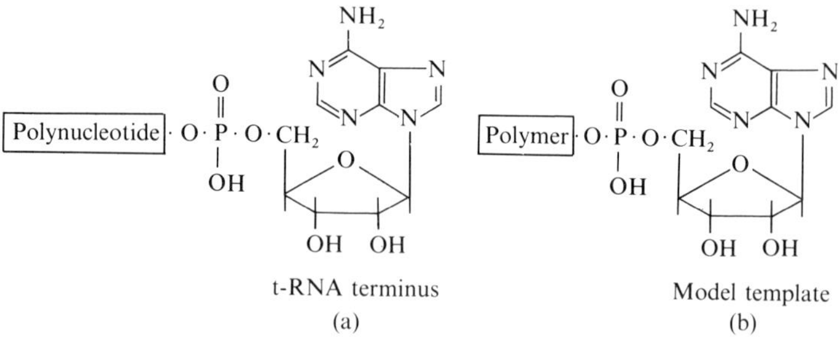

FIG. 8.8. (a) The amino-acid-accepting end of the transfer RNA molecule. (b) The 'model' structure on which the experiments have been carried out.

using as a crude (and far from complete) model of t-RNA, a polystyrene resin with a nucleotide hooked on to it via the phosphate linkage (Fig. 8.8).[21] Exactly the same kinds of coupling experiments were done on this as on the pendent amino-acid system. Instead of coupling amino acid to amino acid, we are now testing the relative efficiency of adding a protected amino acid to a pendent nucleotide sugar to see whether the efficiency of coupling is affected by the nature of the base (Fig. 8.9).[22] Obviously, we can go further. We can build the terminal chain up into a doublet or triplet, having determined the selectivity of the four singlet bases first. The 4 singlets, 16 doublets, and 64 triplets can all be tested in this manner for their ability to accept an amino acid. (This is a long-term experimental programme that is just beginning at Berkeley.)

Figs. 8.7 (a) and 8.9–8.12 show the development of this type of experiment in some detail. Fig. 8.7 (b) shows the two linear polymers for which we are attempting to invent coupling schemes. Fig. 8.9 shows the coupling of the first base to the polymer; Figs. 8.10 and 8.11 show two different methods for the coupling of the amino acids to the nucleotide hydroxyls. The ester link can be on the 2' or 3' carbon. Fig. 8.12 shows the removal of the coupled product for identification. The adenine base with the coupled amino acid first appears on the resin from which it is removed. This material can be synthesized by a completely separate route, using Khorana's method, with triphenylcarbinol in the place of the polymer (polystyrene); the final product is easily removed from this for identification and can be shown to be the same as the product obtained from the resin.

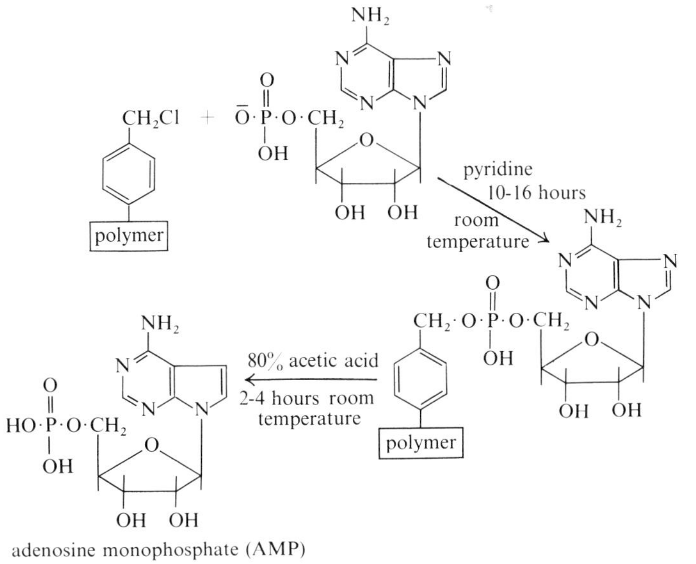

adenosine monophosphate (AMP)

Fig. 8.9. The coupling of AMP with the polymer and the release of free AMP by hydrolysis.

In a qualitative way we have already seen a selectivity of bases for amino acids, the first data for which are shown in Table 8.2. I predict that this selectivity will be widespread, even for a singlet base acceptor, and will be even greater when two bases or more are participating. Whether the spread will be big enough to give the beginning of a coupling system for the two linear polymers remains to be determined in the future.

We have here outlined a purely chemical system that could give rise first to the two linear replicating systems of the polynucleotides and polypeptides, and finally to a coupling of the information in the two systems, making use of the fidelity of the polynucleotide replication and the catalytic ability of the polypeptide system. Once those two properties are coupled it is quite obvious that any regenerative system that had this advantage over the two separate systems will absorb all the available raw materials on to itself.†

We now have the monomers, the replicating linear polymers, information generation, and coupling of information generation. All these processes are not enough. The high efficiency in all of these stages of the

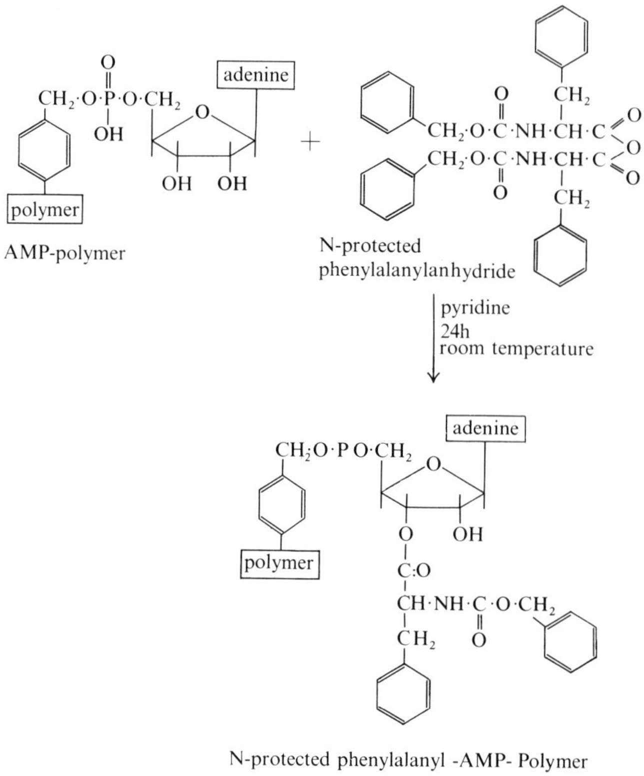

adenine
polymer
AMP-polymer
N-protected
phenylalanylanhydride
pyridine
24h
room temperature
adenine
polymer
N-protected phenylalanyl -AMP- Polymer

FIG. 8.10. The coupling of the polymer-AMP complex with the anhydride form of an N-protected amino acid.

polymer formation, the replication system, and the coupling of the replication system will depend upon ordered chemistry. This chemistry is different from what we are accustomed to thinking of when we put reactive reagents together in a flask, all of them randomly distributed as

† There are those who feel that some additional physical or chemical principles must be called upon to understand this persistence of such an hereditary system. For example, see PATTEE, H. H., The physical basis of coding and reliability in biological evolution. *Towards a theoretical biology*. 1. Prolegomena (editor, C. H. Waddington), pp. 67–94. Edinburgh University Press (1968).

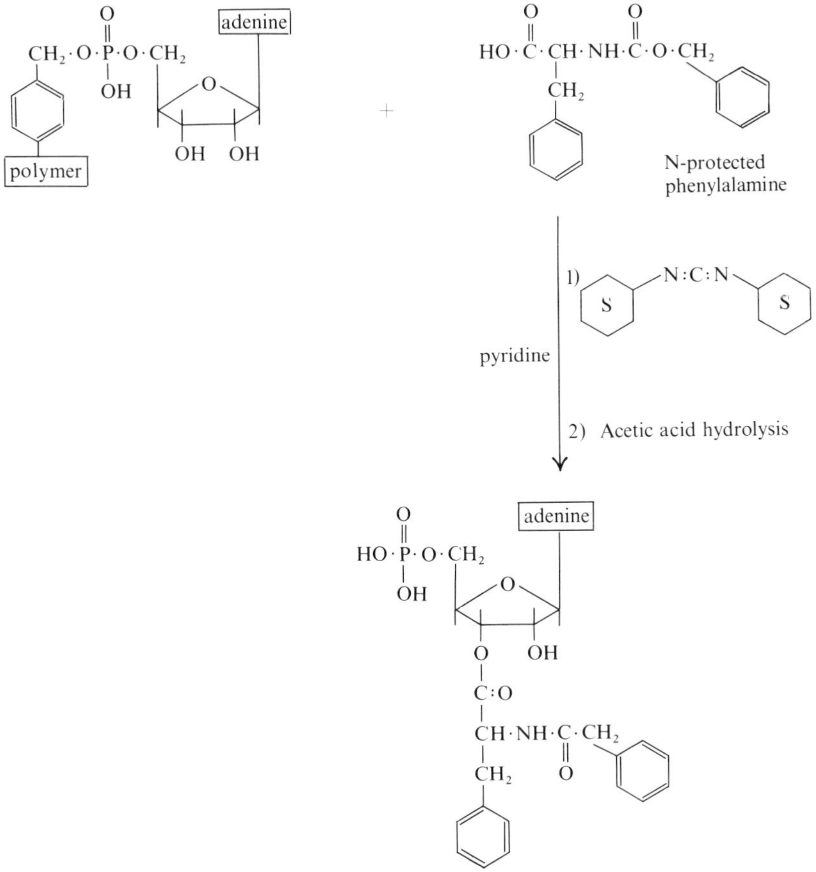

FIG. 8.11. Another method of coupling an N-protected amino acid on to the polymer-AMP complex.

TABLE 8.2

Experiments on selectivity of bases for amino acids. Percentages of bound nucleotide reacting with amino acid.†

	Adenine	Cytosine
Phenylalanine	6·7	2·9
Glycine	10·0	6·5

† (Z–a.a.)$_2$O reacts with polymer-bound nucleotide in pyridine at room temperature for 24 h. The reaction is shown in Fig. 8.9.

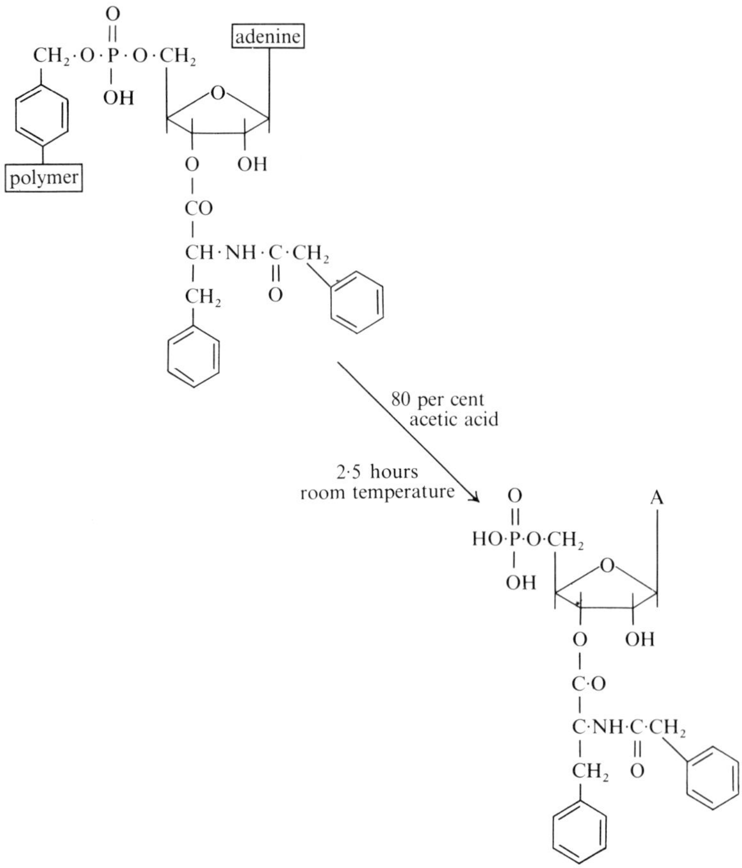

FIG. 8.12. The hydrolysis of the complex to release N-protected aminoacyl t-RNA.

individual molecules. This is what I mean by 'statistical chemistry', as opposed to a chemistry that is done on an ordered structure, the kind of chemistry we are beginning to do with these polymers in using them as matrices for the study of the efficiency of coupling experiments.

REFERENCES

1. For review of dehydration condensation reactions and the chemistry involved, see STEINMAN, GARY, Ph.D. thesis, University of California, Berkeley, December 1965. Also STEINMAN, G., KENYON, D. H., and CALVIN, M., Dehydration condensation in aqueous solutions. *Nature, Lond.* **206,** 707 (1965).
2. STEINMAN, G., LEMMON, R. M., and CALVIN, MELVIN, Cyanamide: a possible key compound in chemical evolution. *Proc. natn. Acad. Sci. U.S.A.* **52,** 27 (1964).

3. SCHIMPL, A., LEMMON, R. M., and CALVIN, M., Formation of cyanamide under 'primitive earth' conditions. *Science* **147**, 148 (1965).
4. STEINMAN, G., LEMMON, R. M., and CALVIN, M., Dicyandiamide: possible role in peptide synthesis during chemical evolution. Ibid. 574 (1965).
5. MILLER, S. L. and PARRIS, M., Synthesis of pyrophosphate under primitive earth conditions. *Nature, Lond.* **204**, 1248 (1964).
6. PONNAMPERUMA, C. A., SAGAN, CARL, and MARINER, RUTH, Synthesis of adenosine triphosphate under possible primitive earth conditions. *Nature, Lond.* **199**, 222 (1963).
7. BARLTROP, J. A., Unpublished results, Chemical Biodynamics Laboratory.
8. KENYON, D. H., STEINMAN, G., and CALVIN, M., The mechanism and proto-biochemical relevance of dicyanamide-mediated peptide synthesis. *Biochim. biophys. Acta* **124**, 339 (1966).
9. STEINMAN, G., and COLE, M. N., Synthesis of biologically pertinent peptides under possible primordial conditions. *Proc. natn. Acad. Sci. U.S.A.* **58**, 735 (1967).
10. ECK, R. V., and DAYHOFF, M. O., *Atlas of protein structure and function*, 1966. National Biomedical Research Foundation, Silver Spring, Maryland (1967).
11. HARADA, K., and FOX, S. W., Thermal polycondensation of free amino acids with polyphosphoric acid. *The origins of prebiological systems and of their molecular matrices* (editor S. W. Fox), p. 289. Academic Press, New York (1965).
12. ITO, EIJI, and STROMINGER, J. L., Enzymic synthesis of the peptide in a uridine nucleotide from *Staphylococcus aureus*. *J. biol. Chem.* **325**, Prelim. Comm. (1960).
13. TSURUTA, T., INOUE, S., and MATSUURA, K., Asymmetric selection in the copolymerization of N-carboxyl-L- and D-alanine anhydride. *Biopolymers* **5**, 313 (1967).
14. MATSUURA, K., INOUE, S., and TSURUTA, T., Asymmetric selection in the copolymerization of N-carboxyl-L- and D-alanine anhydride. *Makromolek. Chem.* **85**, 284 (1965).
15. For a discussion of chain initiation reactions, see *Nature, Lond.* **214**, 759 (1967).
16. SCHRAMM, G., GROTSCH, W., and POLLMAN, W., Nonenzymatic synthesis of polysaccharides, nucleosides, nucleic acids and the origin of self-reproducing systems. *Angew. Chem.* int. Edn. **1**, 1 (1962).
17. NAYLOR, R., and GILHAM, P. H., Study of interactions and reactions of oligo-nucleotides in aqueous solution. *Biochemistry* **8**, 2722 (1966).
18. CALVIN, M., Abiogenic information coupling between nucleic acid and protein. Submitted to *Proc. R. Soc. Edinb.*
19. BERNAL, J. D., The problem of stages in biopoesis. *The origin of life on the earth* (editor A. I. Oparin), pp. 38–53. Pergamon Press, London (1959).
20. CAIRNS-SMITH, A. G., The origin of life and the nature of the primitive gene. *J. theoret. Biol.* **10**, 53 (1966).
21. HARPOLD, M. A., and CALVIN, M., AMP on an insoluble solid support. *Nature, Lond.* **219**, 486 (1968).
22. —— —— Amino acid–nucleotide interactions on an insoluble solid support. I. A simple model of the amino acid acceptor terminus of a t-RNA. (In press.)

9

THREE-DIMENSIONAL STRUCTURE AND SELF-ASSEMBLY

To emphasize the fact that the evolution of such structural features as are required for the highly efficient coupling of energy and information-transferring apparatus is in fact possible, we shall examine some experiments in which these structural features are shown to be intrinsic to the polymers concerned. For example, one does not need a 'Maxwell's demon' to come along and fold up the protein chain in just the right fashion to make it into a catalyst. One does not need the demon to coil up polynucleotides in just the right way for them to perform their functions. Both of these things are endemic, built into the linear sequence itself.

Polypeptide structure

We speak of the linear sequence of units as the primary sequence, or the primary structure, of the polymer. The secondary structure is derived from that primary structure and is built into it, or is a result of it. The first kind of polymer with which we shall deal is the polypeptide itself, shown in Fig. 9.1,[1] which is composed of a sequence of amino acids, with a suitable variety of some twenty different groups on the carbon atoms marked *. The polypeptide sequence might simply be like a loose piece of string, a random coil, and there are circumstances in which this characteristic looseness can be achieved. However, under physiological conditions the coil does not usually randomize but, in a simple case, assumes a regular secondary structure, which is due to the peculiar arrangement of the carbonyl and NH groups along the linear chain, plus the effects of the substituents on the starred carbon atoms. The two extremes of configuration of these chains are a random coil on the one hand, and a helical structure (the α-helix as Pauling named it) on the other, and these two structures are thermodynamically reversibly inter-convertible. For example, a pure homopolymer made of glutamic acid can be reversibly changed from one form to the other by simply adjusting the acidity of the solution, and thus the degree of ionization of the carboxyl side chains on the glutamic acid. The results of such an experi-ment are shown in Fig. 9.2. On the side chain of the glutamic acid is

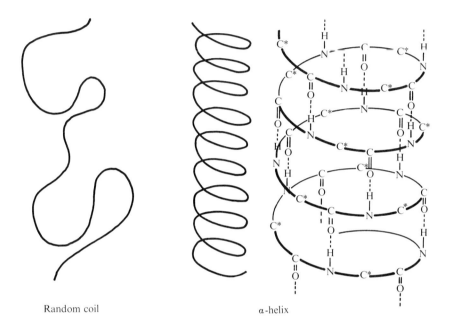

Random coil   α-helix

FIG. 9.1. Secondary protein structure.

a carboxyl group, and if this is ionized, by making the solution slightly alkaline to produce a negative charge on each side chain, the negative charges on the side chains will repel each other electrostatically and prevent any close order of the system, thus resulting in the random coil form. If the carboxyl groups are neutralized by acidifying the solution, the helical structure will re-form. By using optical absorption methods, we can show that at pH 8 we have a random coil, but if the pH is lowered to 4·9 the helical structure results. A single excited state in the random coil becomes a doublet in the ordered helix, and this shows well in the absorption of ultraviolet light. This light absorption provides a means of determining whether we have helical or coiled structure in the polyglutamic acid, and it is quite clear that the molecule can go back and forth very easily. This aspect of the conformation of the chain is the *secondary structure* of the polypeptide chain and it is reversibly achieved as a result of the internal structure of the polypeptide.

What happens to the polypeptide chain in the third stage of its structure, the higher degree of order that can be achieved beyond the helix, is the *tertiary structure*. Each of these structures—primary, secondary, tertiary—is thermodynamically determined by the internal structure of

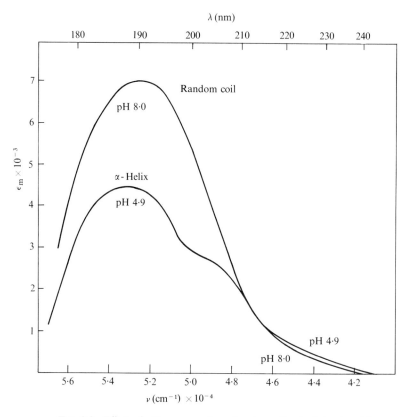

FIG. 9.2. Effect of pH on secondary structure of polyglutamic acid.

the material: the primary determines the secondary, and the secondary the tertiary. We shall briefly examine these phenomena with experimental examples in each case.

Fig. 9.3 shows the structure of the myoglobin molecule, which has a tertiary structure containing in it some of the elements of secondary structure, i.e. the helical components as well as the bends in the non-helical parts of the chain.[2,3] This is what we mean by tertiary structure: the folding into a specific configuration of a macromolecule containing a particular sequence of units, with a specific amount of secondary structure, all of which gives rise to a specific tertiary structure. Tertiary structure is a stable configuration that can be removed and reconstructed in

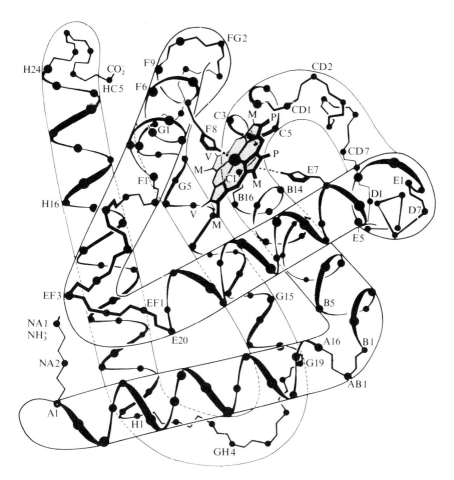

FIG. 9.3. The structure of the myoglobin molecule, showing the several straight sections of alpha-helix linked by short lengths of non-helical chain. The haem group, with an iron atom at its centre, is held in place between the chains.[3]

the same way that the secondary structure is a stable configuration of the linear array that can be destroyed and re-formed under suitable changes of outside conditions. I have data for one molecular example of this phenomenon, which I should like to review here. Fig. 9.4 shows the change in extinction coefficient as a function of temperature for chymotrypsin, a more complex protein molecule. The observation is on the absorption of light in the region of 290 nm (2900 Å), outside the absorption region characteristic for the helical structure; this absorption is the result of the tertiary folding, not the direct result of the secondary structure.[4] One can see in Fig. 9.4 the melting out or disappearance of the tertiary folding as a function of temperature at several pH values. At pH 3, note that between 5 and 35 °C there is no change whatsoever in the shape of the protein molecule; above 40 °C, it 'snaps' or 'melts' out of its tertiary structure into an open structure, but the helix

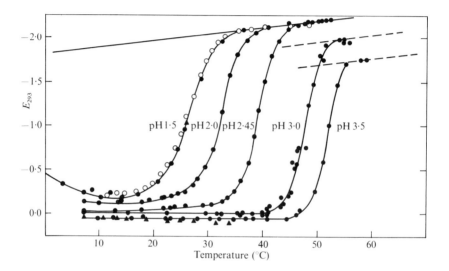

FIG. 9.4. Change in extinction coefficient as a function of temperature for α-chymotrypsin at various pH values.[4]

remains. The evidence for the fact that the helical structure in the chymotrypsin has not been destroyed is that the optical rotation of the molecule in the wavelength region below 220 nm (2200 Å) is unchanged, or relatively little changed. This change is reversible, and upon cooling the molecule goes back into its original tertiary folded configuration. We can thus be confident that the tertiary folding of chymotrypsin represented in this transformation is recoverable upon cooling. Chymotrypsin has less helical structure in it than myoglobin; its conformation is shown in Fig. 9.5. The melting behaviour of the chymotrypsin described above is the unfolding of the tertiary structure and not of the secondary helix, and upon cooling it is refolded into the normal configuration.

Still another way of demonstrating this change in tertiary structure has been achieved with a third protein molecule, lysozyme (Fig. 9.6).[6]

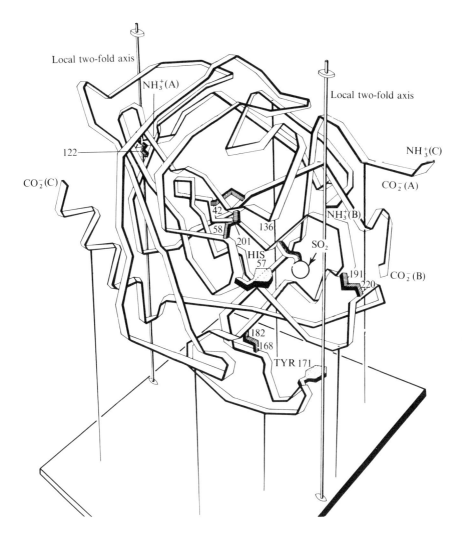

FIG. 9.5. The conformation of the chymotrypsin molecule, which consists of three separate chains, A, B, and C, held together by disulphide bridges. The only alpha-helix is near one end of the C-chain. (Drawing by Miss Annette Snazle.)

There is considerably more helical content in the lysozyme, as indicated by the cross-hatched parts. A quite different technique has been used to examine the tertiary changes in lysozyme: the change in the NMR spectrum—the spectrum of the protons, particularly of those in the various side chains of the amino acids in the structure, as a function of temperature. The results of this examination are shown in Fig. 9.7.[5] The change in the proton peaks in the lysozyme, in going from the folded condition at 51 °C to the unfolded condition at 76 °C, shows that the helix is not melted out; this, again, is a reversible change. NMR analyses can give

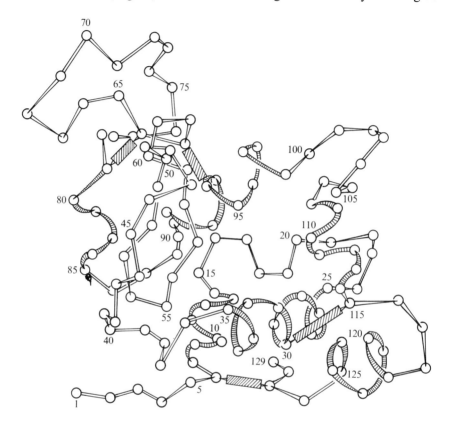

FIG. 9.6. Sketch by Sir Lawrence Bragg of the lysozyme molecule. The relatively short sections of alpha-helix are indicated by shading. The shaded bars are disulphide bridges linking parts of the polypeptide chain.

us the most intimate information about the shape and interaction of such molecules in solution in a manner that will make it possible for us to understand a little more of the causes of this tertiary structure.

Polynucleotide structure

With the polypeptides we have traced a sequence of organization from the sequence of amino acids giving primary structure, through the helical form that arises as a result of the stringing together of amino acids and

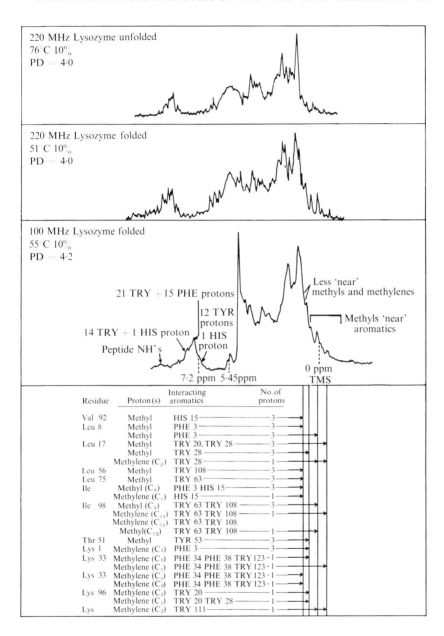

FIG. 9.7. Changes in NMR spectrum of lysozyme with temperature, indicating (reversible) unfolding of the helical structure.[5]

of the environmental conditions, to a tertiary structure that is the result of the arrangement of amino acids in any particular sequence. The detailed factors determining that tertiary structure are yet to be completely unravelled.

For the other linear sequence, the polynucleotides, a similar set of observations can be described to show that here again there is a primary structure (the linear sequence) and a secondary structure (the helical conformation). Whether or not there is a tertiary structure analogous to that of the polypeptides is not yet clear. There is a 'super-coiling' effect in the polynucleotides, but it has not yet been as clearly defined.

To give some deductive arguments that would lead to the formation of the secondary structure of the polynucleotides, we must discuss another kind of force that acts here, which is quite different from the one primarily responsible for the appearance of secondary structure in the polypeptides. We recall that the secondary structure in the polypeptides is the result, in the first instance, of hydrogen bonding between the NH group of the peptide linkage and a carbonyl group of another peptide linkage, several bonds removed. This tends to give rise to a helical configuration. The variety of side-chain interactions in the polypeptides is also of great importance in stabilizing this helical formation. The forces in the polynucleotides that tend to create the secondary structure are of a different character. These forces are, to a first approximation, the $\pi-\pi$ interactions of the aromatic rings of the polynucleotides with each other in a stacking configuration. Such interactions have long been known in organic chemistry. They were described as long ago as the 1930s for a variety of dyestuffs in solution. This stacking phenomenon was discovered quite independently by two workers, Scheibe[6] in Germany and Jelley[7] in the United States, in connection with their studies on the physical state of cyanine dyes in solution. The cyanine dyes are large, flat molecules, all of which have positive charges on them and simple negative anions accompanying them. The dyestuffs tend to orient themselves, even in very dilute solution, so that the flat molecules lie one on top of the other. The experimental evidence came largely from optical absorption. Fig. 9.8 shows how the flat molecules stack upon each other; the distances between planes here are rather large. This effect can also occur in exactly the kind of structure that hydrogen bonding produces among the base pairs of the nucleic acid bases strung together along the sugar phosphate chain (Fig. 9.9). When the pair of sugar phosphate chains (a double linear array) is twisted, the aromatic groups will turn around face to face, one on top of the other. Fig. 9.10 shows the classic

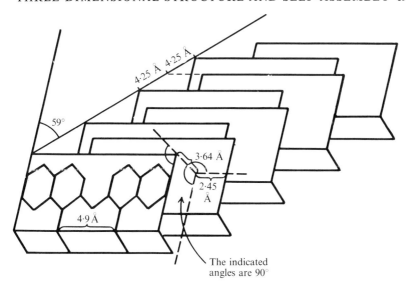

FIG. 9.8. Packing arrangement of a cyanine dye in solution, showing the arrangement of the plate-like molecules of pseudo-isocyanin when they associate.[6]

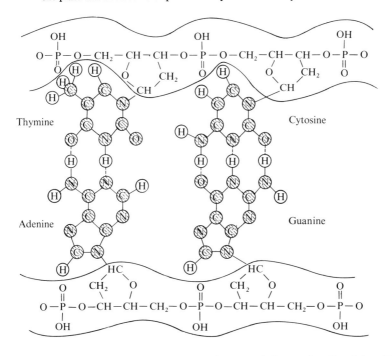

FIG. 9.9. Hydrogen bonding between aromatic groups in base pairs of nucleic acid bases and twisted sugar-phosphate chain.

picture of the double helix of the DNA chain, developed by Watson and Crick in 1953.[8,9] The horizontal bars in the twisted chain represent the aromatic double base pairs, sitting one on top of the other, roughly 0·34 nm (3·4 Å) apart. Therefore, built into the primary sequence of the

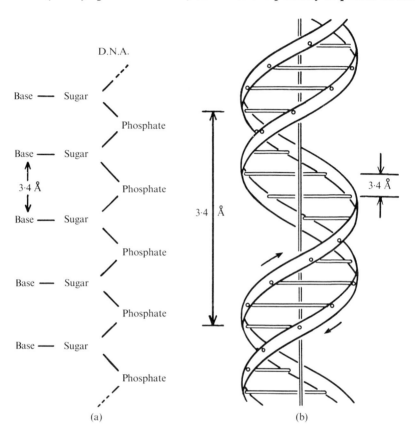

(a)   (b)

FIG. 9.10. (a) Chemical formula of a single chain of deoxyribonucleic acid. (b) Purely diagrammatic figure in which the two ribbons symbolize the two phosphate-sugar chains, and the horizontal rods the pairs of bases holding the chains together; the vertical line marks the fibre axis.[8]

polynucleotides is this stacking helical arrangement, just as the helical arrangement is built into the polypeptides.[10]

This can be demonstrated experimentally by going through a temperature sequence that melts out and destroys the ordered array in the double helix. If the temperature is high enough, a change in the optical properties of the stacked series of bases can be observed; when the material is cooled, it returns to its original condition. This effect is shown in Fig. 9.11, which gives the absorption spectrum of a helical form of native DNA; at 99 °C the helix is melted out, with a consequent change in absorption spectra.[1] One can use this difference to measure melting temperatures of such large molecules; by that I mean the temperatures at which the secondary structure vanishes. This structure is thermo-dynamically stable and forms reversibly in much the same way as the

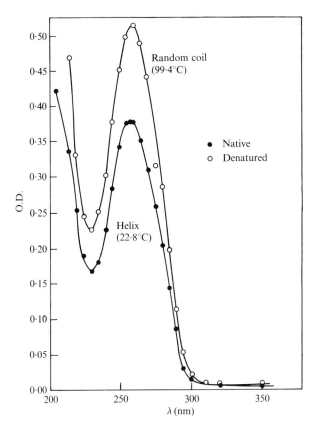

FIG. 9.11. Absorption spectra of a helical form of native DNA, showing effect of temperature on the helix.

secondary and tertiary structure of the polypeptide. Thus the primary sequence of the polynucleotides contains in it higher degrees of order, to the secondary level. For the polypeptides the higher degrees of order go through the tertiary as well. While there is some experimental evidence for tertiary structure in nucleic acids, particularly among the transfer-RNA molecules, it is not yet sufficiently well developed to be included here.

Quaternary, and higher orders of structure

The development of tertiary structure is only a step in the development of structural features in living organisms. The next level of structure, in which the folded helices of individual covalently linked chains begin to aggregate into some higher degree of order, has been demonstrated in a variety of situations, which I shall describe in order of increasing degree of complexity. The first is the self-assembly of relatively simple protein molecules, such as the simple single protein molecules contained in a collagen fibril; this is one of the earliest examples of reassembly of which I became aware. Collagen is one of the connective-tissue proteins, but not a very complex one in so far as any chemical functional variability is concerned; it is really a structural protein. This simple self-assembly process was demonstrated some years ago.[11] Fig. 9.12 (b) shows the complex fibrils made of many collagen molecules; Fig. 9.12 (a) shows the result of disassembling the collection of molecules into individual protein molecules: these individual strands of the single protein molecules are stretched out in the photograph. If the salt concentration and pH of the medium are adjusted, these individual protein molecules reassemble into the collagin fibrils, virtually identical with the original. This is not a very complex reassembly process, for collagen is a rather simple molecule with a relatively simple quaternary structure, a linear array of protein molecules.

I shall now describe the self-reassembly of an enzyme, which has chemical function, rather than just structural function. This is a more difficult process. The enzyme in question is transacetylase, the enzyme that will take the acetyl group from the sulphur atom of reduced lipoic acid and transfer it to the sulphydryl group of coenzyme A. This enzyme is a multiple molecule in its intact form; it is made up, not of one protein molecule, but of 32 identical protein molecules, each with a molecular weight of about 35 000. When the molecule of transacetylase is isolated in its native form, it has a molecular weight of $32 \times 35\,000$. However, it can be dissociated into its individual constituents in stages.[12] The active enzyme can be dissociated into a dimer by isolation in acetic acid; this dimer is not enzymatically active. One can also obtain the monomer. Finally the transacetylase can be reassembled into its complete active form. Although the monomer and dimer are not enzymatically active, the reassembled enzyme is. The assembled enzyme has a very characteristic cubic structure. Electron micrographs show it as a cube, as though there were groups of molecules at the eight corners of a cube

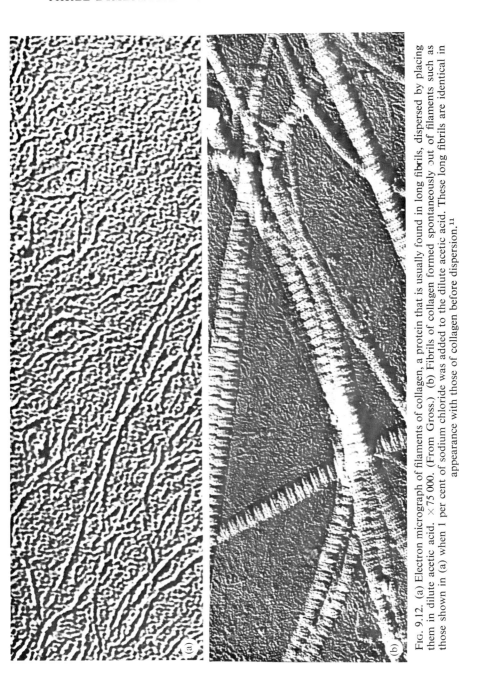

FIG. 9.12. (a) Electron micrograph of filaments of collagen, a protein that is usually found in long fibrils, dispersed by placing them in dilute acetic acid. ×75 000. (From Gross.) (b) Fibrils of collagen formed spontaneously out of filaments such as those shown in (a) when 1 per cent of sodium chloride was added to the dilute acetic acid. These long fibrils are identical in appearance with those of collagen before dispersion.[11]

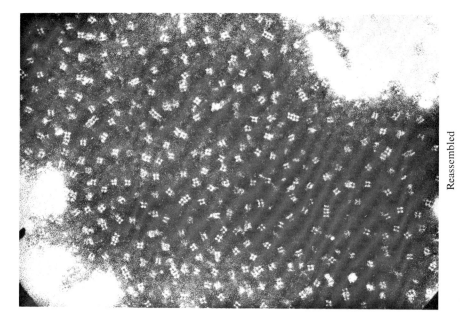

Reassembled

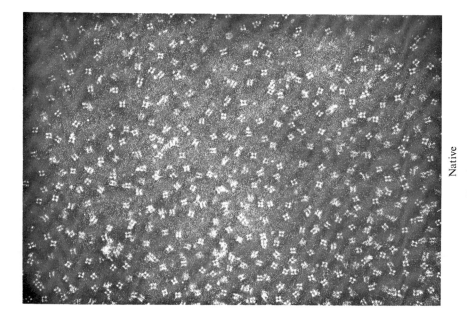

Native

FIG. 9.13. Electron micrograph of 'native' and 'reassembled' transacetylase.[12]

(Fig. 9.13), each one being a tetramer (a dimer of a dimer); there is therefore a high degree of symmetry. The four corners of one face of the cube are easily seen in the figure, and the molecules all show the same features; on the right-hand side of the figure is the 'native' enzyme and on the left the reassembled enzyme; they are virtually identical. There is also some suggestion that the tetramers of this molecule might stack into something larger than a cube, but the experimental evidence for this is not yet definitely established.

Next we shall examine a still more complex assembly process in which we see not only reassembly into chemically active forms but selective reassembly from a mixture—the molecules of the same enzyme from two different organisms, even of the same phylum, can tell each other apart. The molecules concerned are haemocyanin, the copper-containing enzyme, which corresponds to haemoglobin for the phyla of the molluscs and the arthropods, at least. These rather large molecules are assemblies of varying numbers of identical units, from 100 to 1000, according to the particular species of snail or crab. Haemocyanin is a classic example of the assembly of identical units and was one of the first of the proteins to be studied by Svedberg in his ultracentrifuge 30 years ago. Recently the question of self-assembly and selective self-assembly has been re-examined in the electron microscope by Fernandez-Moran at Chicago, and Figs. 9.14–9.23 are a selection of some of the micrographs made there[13]. Fernandez-Moran was able to prepare the haemocyanin from *Helix* (a kind of snail) and *Loligo* (another snail) and from *Limulus* (an arthropod, a crab). These various haemocyanins are characterized by different numbers of subunits in their final structure, and the final structure in each case has a characteristic shape that is easily recognizable in the electron microscope. Figs. 9.14–9.18 show examples of haemocyanins from organisms belonging to different phyla, which clearly demonstrate selective reassembly across different phyla. The next group, Figs. 9.19–9.23, shows the selectivity in reassembly processes between corresponding molecules of different organisms within the same phylum; the haemocyanins of *Helix* and *Loligo*. Fig. 9.21 shows the intact molecules of haemocyanin of *Helix* and *Loligo* mixed together; (b) and (c) show the *Helix* haemocyanins, which are cylindrical in shape as are those of *Loligo* (d) and (e), but they are of a different height and a slightly larger diameter, and the interior of the *Loligo* is somewhat more readily stained. The *Helix* and *Loligo* haemocyanins can be recognized in the mixture of intact molecules of Fig. 9.21 (a). Fig. 9.22 shows the mixture of the disaggregated haemocyanins of *Helix* and *Loligo*; the molecular weight of

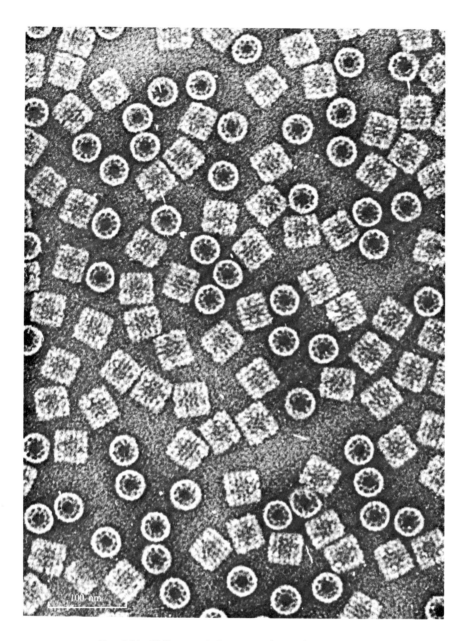

FIG. 9.14. *Helix pomatia* haemocyanin uranium EDTA.[13]

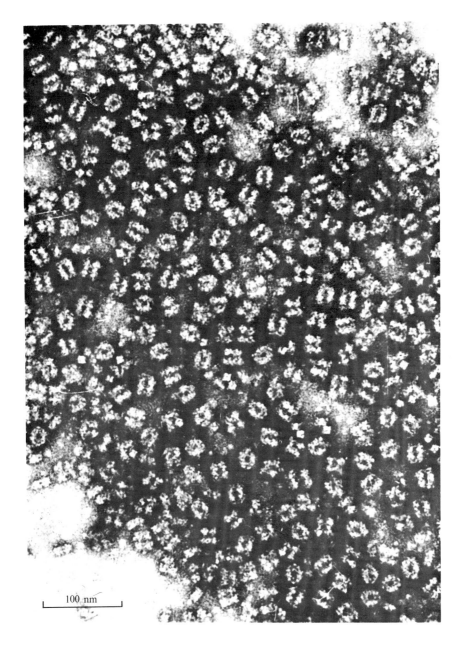

100 nm

FIG. 9.15. *Limulus polyhemus* haemocyanin sodium phosphotungstate, showing four components.[13]

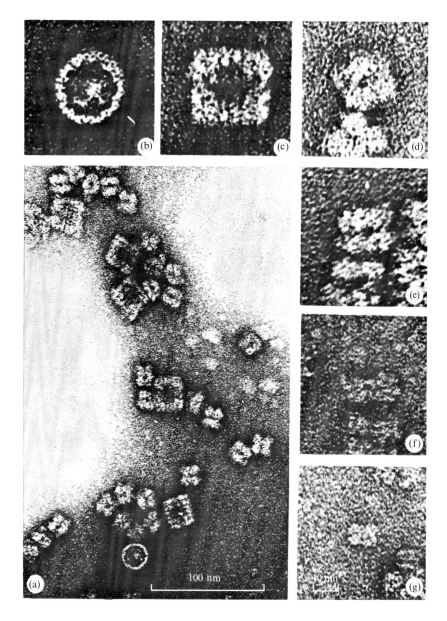

FIG. 9.16. Mixture of *H. pomatia* and *Limulus polyhemus* haemocyanin. (b, c) *Helix*, (d–g) *Limulus*.[13]

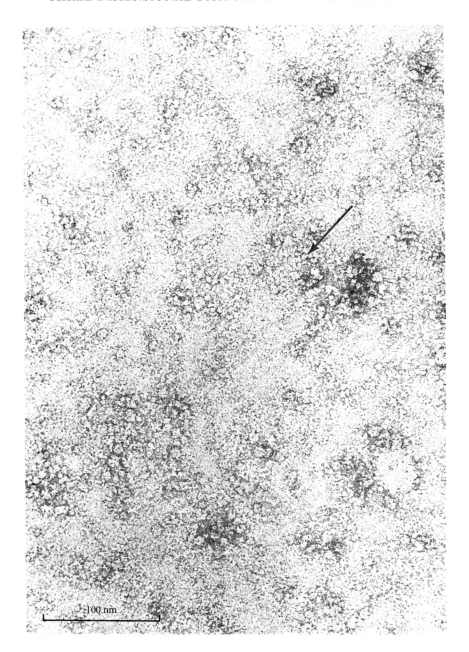

FIG. 9.17. Mixture of *Helix* and *Limulus* haemocyanin subunits.[13]

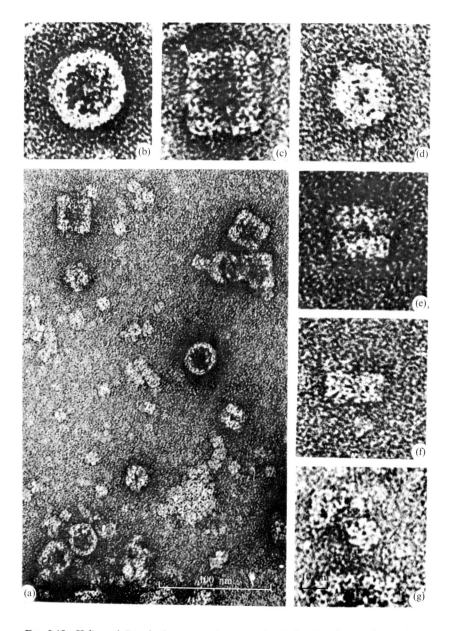

FIG. 9.18. *Helix* and *Limulus* haemocyanin, reassociated. (b, c) *Helix*, (d–g) *Limulus*.[13]

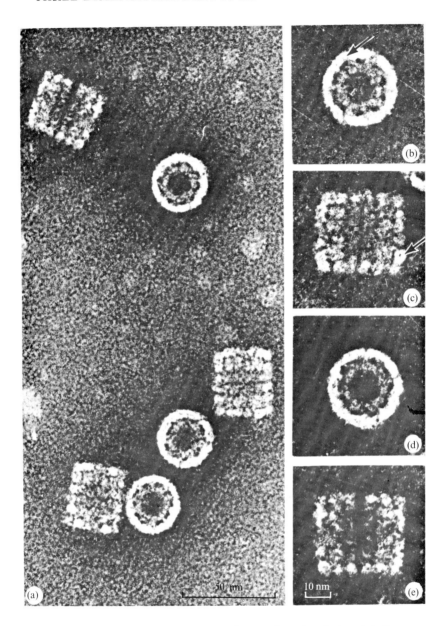

FIG. 9.19. *Helix pomatia* haemocyanin potassium phosphotungstate.[13]

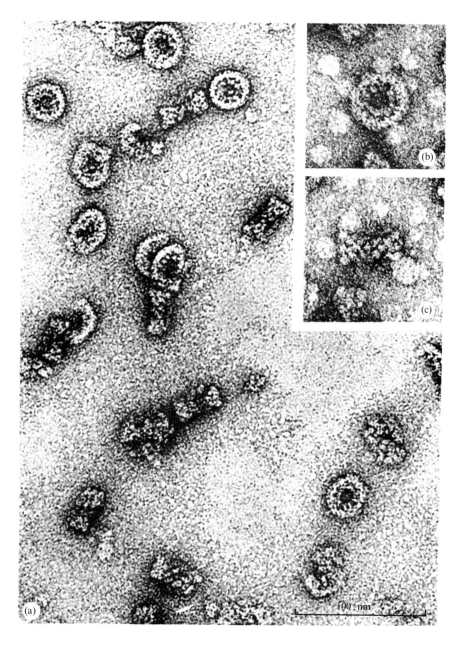

FIG. 9.20. *Loligo pealei* haemocyanin uranium EDTA.[13]

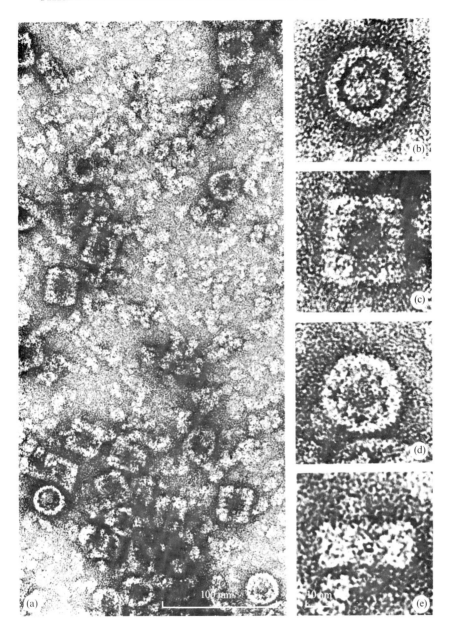

FIG. 9.21. *Helix* and *Loligo* haemocyanin. (b, c) *Helix*. (d, e) *Loligo*.[13]

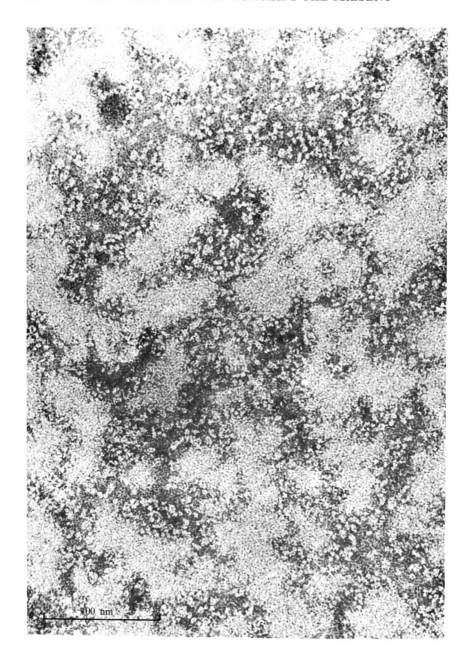

FIG. 9.22. Mixture of *Helix* and *Loligo* haemocyanin subunits.[13]

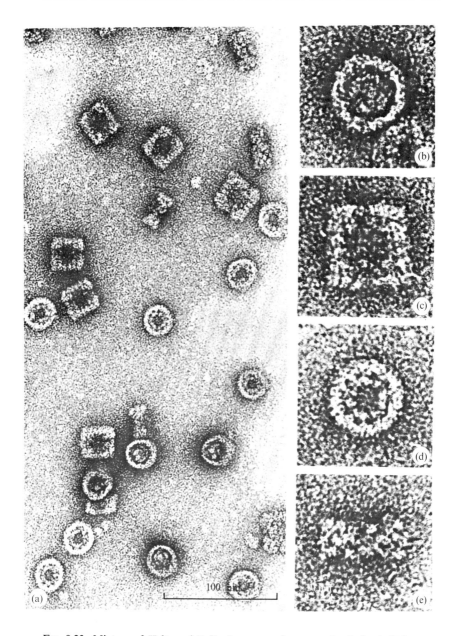

FIG. 9.23. Mixture of *Helix* and *Loligo* haemocyanin, reassociated. (b, c) *Helix*, (d, e) *Loligo*.[13]

the subunits is only about 35 000 and they are very hard to observe. Fig. 9.23 shows the result of the reassociation process; two molecules (of *Helix* and of *Loligo*) reassemble separately. These figures illustrate a selective process of reassembly that clearly must be built into the primary sequence. I have not given here the primary sequences of the two different haemocyanins, partly because we do not yet know them in sufficient detail. Undoubtedly there will be differences in the two primary sequences, and eventually we shall be able to deduce from these differences what are the differences between the subunits that, in turn, give rise to a difference in the reassembly process and recognition ability and keep the two separate subunits from becoming mixed.

We now have a fairly high degree of selectivity in molecular recognition, which we have been able to trace from primary sequence up to something that can be seen in the electron microscope. All the reassembly processes that have been described so far—collagen, transacetylase, haemocyanin—are assemblies of identical subunits. The selectivity process has, however, gone much further than that.

The transacetylase enzyme discussed earlier is really one of a complex of three different enzymes that was originally thought to be a single enzyme, namely pyruvic acid oxidase. Lester Reed was able to show that this pyruvic acid oxidase was actually made up of three different enzymes, of which transacetylase was one; its reaction sequence is shown in Fig. 9.24.[12] The three enzymes are a decarboxylase, a transacetylase (the little cube with the eight corners and eight polypeptide chains at each one of the corners of that cube, shown in Fig. 9.13), and finally the flavin enzyme, dihydrolipoyl dehydrogenase (FAD), which transfers the hydrogen from the reduced lipoic acid to pyridine nucleotide, making it into reduced pyridine nucleotide. Not only do we know that there are three separate enzymes in this particular reaction, but we know that these enzymes can be separated from each other and reassembled selectively into a single functional and structural unit. Fig. 9.25 (a) shows the native enzyme—a very complicated organization but, nevertheless, containing an ordered array. Fig. 9.25 (b) is an electron micrograph of the reassembled pyruvic dehydrogenase, a structure of all three enzymes reassembled from all the subunits of the three separate enzymes. The reassembly mechanism is quite complete and the reassembled enzymes function just as the native complex does. Fig. 9.26 shows the structures of the pyruvic dehydrogenase system. The transacetylase (Fig. 9.26 (b)–(d)) is composed of eight units at the corners of a cube; sometimes the cubes are seen on edge, thus showing the three units.

Fig. 9.26 (g) and (h) shows the model of the completely reassembled pyruvic acid oxidase, composed of three different enzymes. Reed has pictured the pyruvic acid oxidase with the transacetylase cube in the centre; on each of the six faces of the transacetylase cube there is a molecule of the dehydrogenase, and on each of the 12 edges of the cube a molecule of decarboxylase. Thus, this complex enzyme is made up of one cube of transacetylase; six dehydrogenases, one on each face of the cube; and 12 decarboxylases, one on each edge of the cube. In this

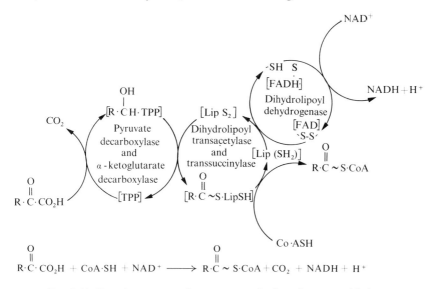

FIG. 9.24. Reaction-sequence in pyruvate and α-ketoglutarate oxidation.

system the reassembly process is much more complex; a reassembly not only of identical subunits to give an 'active' enzyme, but of clearly different kinds of proteins, to give a complex of enzymes. We are now beginning to approach 'chemical systems', the type of thing we must have to give rise ultimately to metabolism and evolution of the kind the biologists are concerned with.

Until now we have been dealing only with chemical structures, the physical structures arising from them, and the self-assembly of these entities to give functional units.

There is one other self-assembling system to which I should like to call attention: the reassembly of a functional structure of a more complex type, a bacterial flagellum. This is certainly a more complicated structure, for it can move by converting chemical into mechanical energy.[14] Fig. 9.27 shows four types of *Salmonella* flagella, each one of

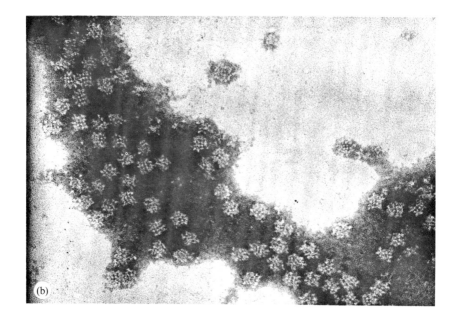

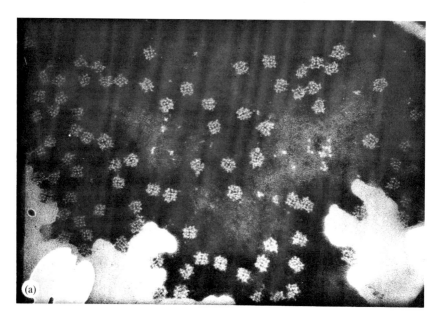

FIG. 9.25. (a) Natural pyruvic dehydrogenase, (b) reassembled pyruvic dehydrogenase.

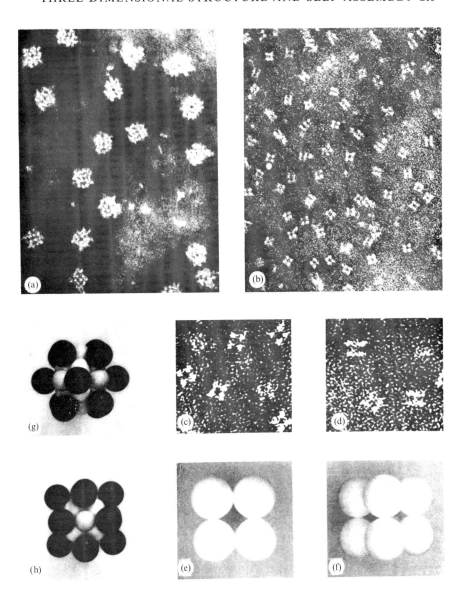

FIG. 9.26. Structures of the pyruvic dehydrogenase system. (a) pyruvic dehydrogenase, (b–d) transacetylase, (e, f) views of the enzyme model corresponding to the detail in (c) and (d) respectively, (g, h) views of the model of the reassembled enzyme system.[12]

the four having a different wavelength in the single flagellum. The tails, which are essentially protein, can be shaken off the bacterial body and can be dissociated into molecules of a protein called flagellin. These molecules can then be reassembled into a flagellum; not just any flagellum, but one of the right wavelength. This fact is, I think, important in terms of specific structure. Note particularly the wavelengths for III and IV in Fig. 9.27, because this information will be used subsequently.

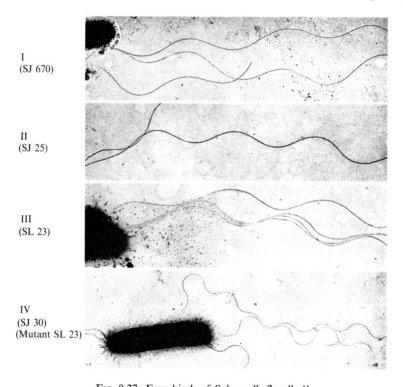

I
(SJ 670)

II
(SJ 25)

III
(SL 23)

IV
(SJ 30)
(Mutant SL 23)

FIG. 9.27. Four kinds of *Salmonella* flagella.[14]

Specific reassembly of each type of *Salmonella* flagella is shown in Fig. 9.28. Reassembly of type III, seeded by a small amount of the end-product, gives flagella III. The monomer of flagella IV was seeded with III, on the assumption that the reassembly was some kind of crystallization process, which it obviously cannot be; although it can be seeded with flagella III, the resultant material has the wavelength of IV. Figs. 9.27 and 9.28 illustrate the reassembly of a very complex material, giving rise to something that has not only a chemical function but a physical function as well; it participates in the conversion of chemical energy into mechanical energy. We now have a self-assembly of such a process, from relatively simple protein molecules.

All the assembly processes so far discussed are reassemblies of protein, but the self-assembly and reassembly process has been carried further.

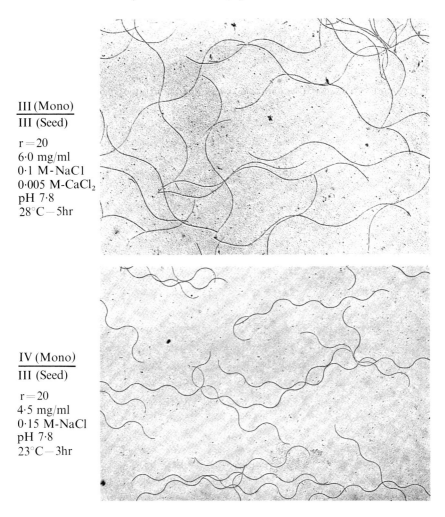

III (Mono)
III (Seed)

r = 20
6·0 mg/ml
0·1 M-NaCl
0·005 M-CaCl$_2$
pH 7·8
28°C—5hr

IV (Mono)
III (Seed)

r = 20
4·5 mg/ml
0·15 M-NaCl
pH 7·8
23°C—3hr

FIG. 9.28. Reconstituted *Salmonella* flagella.[14] (× 15 000)

It has been possible to reassemble not only, as has been outlined, identical subunits, non-identical subunits, proteins of different enzyme composition to a specific array of enzymes; it has been found possible to reassemble two different kinds of polymers, namely, protein and nucleic acid. When we talk about the reassembly of these two different kinds of polymers into a single functional unit in a specific reassembly process, we are really talking about the assembly (or reassembly) of a virus particle, which is composed essentially of protein and nucleic acid. A virus usually has a nucleic acid core, which carries the information for the construction of the virus, and on the outside a protein coat, for protection in the first instance, and to provide the means of entry into the

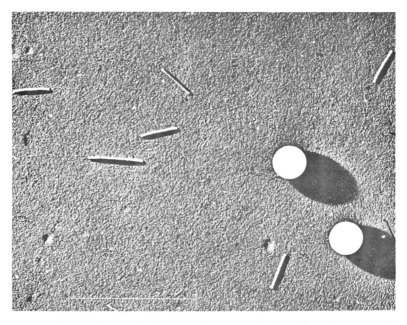

FIG. 9.29. Electron micrograph of native tobacco mosaic virus. The polystyrene spheres (markers) are about 60 nm (600 Å) in diameter.

specific cell that is subject to infection by the virus. The earliest reassembly process of this mixed type was the reassembly of tobacco mosaic virus (TMV). This was done in Berkeley some 10 years ago by Dr. Robley Williams.[15] The virus was disassembled by suitable chemical means; the protein was separated from the nucleic acid. It was then that the nucleic acid component of the virus (a ribonucleic acid component rather than a DNA component) was shown to be enough to infect tobacco plants and make whole viruses grow. The physical assembly of these molecular entities occurs in a specific way, and TMV was the first case of specific reassembly of a visible functioning structure. Figs. 9.29–9.31 show the disassembly and reassembly processes for TMV virus. Fig. 9.29 is an electron micrograph of native TMV; the TMV particles

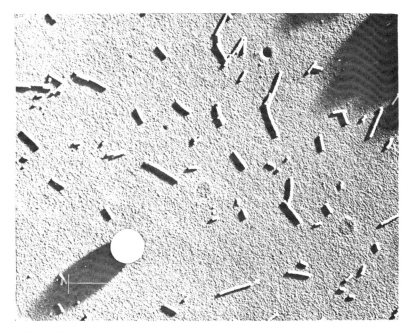

FIG. 9.30. Repolymerized TMV protein (only). Approx. same scale as Fig. 9.29.

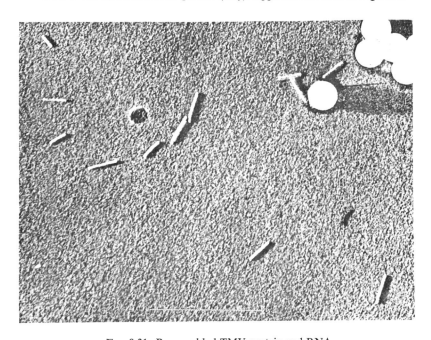

FIG. 9.31. Reassembled TMV protein and RNA.

are rods of fairly uniform length. This TMV can be disassembled into two pieces, a very long strand of RNA and a collection of several hundred identical protein molecules. The ability to disassemble and reassemble has provided us with chemical means of identifying many complex macromolecular substances, and is a very useful tool indeed. Fig. 9.30 shows the reassembly of the protein in the absence of nucleic acid; this also forms a rod-like structure, somewhat similar to the native TMV shown in Fig. 9.29. As the nucleic acid containing the genetic information is missing, the protein itself builds irregularly long rods and short ones, varying considerably in length. The complete programme for reconstruction of the virus is therefore not contained in the protein alone. When the two materials are put together the reassembled virus appears, as shown in Fig. 9.31, containing rods of the proper length. These rods, as far as we can tell, are identical with the native TMV virus. This is a reassembly process, therefore, of two chemically different kinds of material, specifically reassembled to give the proper structure and 'organism'. The construction of what is marginally, if not unequivocally, a living organism is self-contained within the structure of the polymers of which it is made.

Except for the very first bits of secondary structure, these reassemblies of polymers have been obtained from living organisms. The only exception has been the synthetic polypeptides, which, as we can show, assemble reversibly into their secondary helical structure; some of the synthetic polynucleotides can also be shown to adjust their helical structure in a reversible fashion. Only the first stage of assembly—from primary structure to secondary structure—has thus been carried out entirely with synthetic materials.

To conclude this chapter, I should like to describe a step in the complex reassembly of a virus in which at least one part of the virus has been partly synthetic. This is work that has been done with a virus that infects bacteria, the phage ϕX-174. This virus has the appearance of a polyhedron, as shown in Fig. 9.32; there is a variation in the shadow ratio between Fig. 9.32 (a) and Fig. 9.32 (b), which shows the difference in size of the virus. McLean and Hall suggested that the virus particles are regular icosahedra with 20 faces, and 12 pentagonal vertices.[16] Fig. 9.33 shows a diagram of such a particle, with 20 such faces meeting at 12 pentagonal points. The spheres represent protein molecules, and what is seen in the figure is the electron density due to the shadowed protein units. The question of how many proteins go into making up this curiously shaped protein coat of the phage virus is still open. In any case,

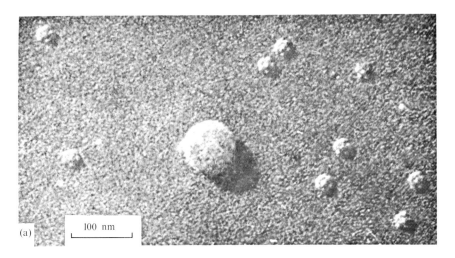

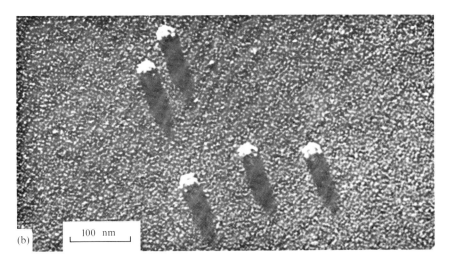

FIG. 9.32. (a) Bacteriophage ϕX-174 shadowed at 1:1, $\times$ 180 000, (b) bacteriophage ϕX-174 shadowed at 5:1, $\times$ 180 000.[16]

the virus nucleic acid is surrounded by the coat of protein. This particular virus has not been carried through the total reassembly process in the same way as TMV has. However, it has been partly reassembled to an infective form. I think it is largely a matter of learning what the proper conditions of disassembly are, so that the specificity in the structure is not destroyed, before the proper conditions of reassembly are found.

Recently the nucleic acid component of this phage virus has been synthesized outside a cell, in a series of test-tube reactions.[17] In that sense

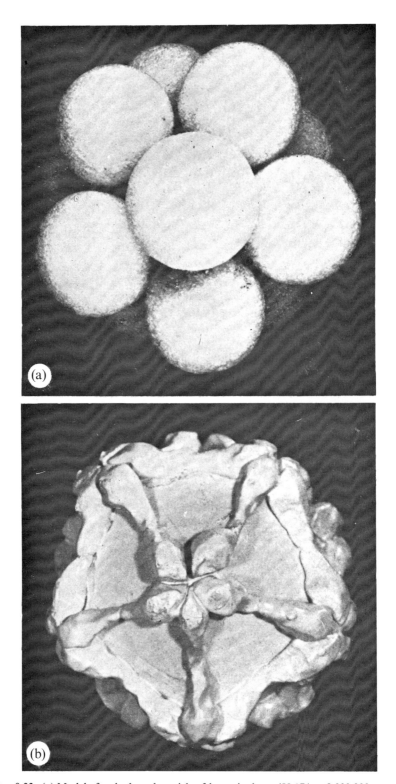

FIG. 9.33. (a) Model of a shadowed particle of bacteriophage ϕX-174, $\times$2 000 000 approx. (b) Model of an unshadowed particle of bacteriophage ϕX-174, $\times$2 000 000 approx.[16]

we have part of the virus particle, now partly synthetic, as shown dia-
grammatically in Fig. 9.34. I should like to outline briefly the sequence
of events that Kornberg and his colleagues went through in order to
reconstruct, *in vitro*, the (DNA) of ϕX-174 and show that it was infective. They used the templating ability of one of the strands of ϕX-174
taken from an existing virus (referred to as the + strand). The DNA
base pairing gives rise to a double strand, but the double strands are not

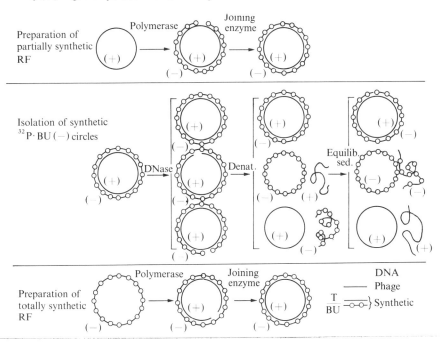

FIG. 9.34. Synthesis of nucleic acid of ϕX-174.[17]

identical since A pairs with T and G pairs with C. Kornberg added DNA
polymerase to a mixture of one of the strands with the four nucleic acid
bases that were presumed to be present in that strand. The bases were
used in the form of their sugar phosphates and some ^{32}P phosphates were
included in the reaction-mixture. Actually, bromouracil was used instead
of one of the nucleic acid pyrimidine bases, thymine. It has a bromine
atom in place of the methyl group of the thymine, and the bromine atom
is very heavy when compared to a methyl group although it does not
differ much in size. The circle was closed by another specific catalyst, and
a circular double helix formed. Kornberg took this circular helix con-

taining the bromouracil and some radioactive phosphorus, and treated it, now separated from all the enzyme and non-usable material, with another enzyme to break a phosphate ester linkage randomly, on both the inside and outside chains (+ and − chains). By adjusting the concentration of the enzyme so that not too much of the virus nucleic acid is destroyed, some of the intact double circles are left, and there are also some with the inside circle cut and some with the outside one cut. If the mixture of all three types is heated at the right temperature and for the right length of time, the double circle remains intact while any broken chains will come off the intact ones. A mixture of five things will result: double circles, circular heavy ones, circular light ones, the denatured light chains, and the denatured heavy chains. The sedimentation characteristics of these five materials are all different, and by equilibrium sedimentation the synthetic circular heavy nucleic acid chain (−) can be separated from the other four components. The circular heavy chain, which was made synthetically, can now be used again as a template, again in a test-tube, and the operation repeated, this time to make a completely 'synthetic' light (+) chain. Again, the same operation can be repeated; one can fracture some of these light chains and get a totally synthetic separate (+) chain. Kornberg showed that the totally synthetic (+) chain is indeed infective and can produce a completed virus in *E. coli*.

Thus, we have carried through the whole sequence of events from the monomer through the polymer of both types (polypeptide and polynucleotide), and have indicated a manner by which these two linear sequences can be coupled, informationally. We have seen how the two linear sequences can grow and associate into visible and functional structures. Such a system is not yet a living cell, and is probably some distance from being one. We still have some steps to demonstrate—and there are those who feel that this is one of the most important features—the separation of these entities from the world at large. We still have to generate a local concentration of materials and separate it by some kind of semipermeable barrier, some kind of a *membrane*, if you like, that will define the unit and separate it from the mixture of all the molecules present in the primitive ocean.

In the next chapter we shall examine this, the most difficult process and perhaps the least founded on experiment. We shall determine how far we can get, which natural processes that have an experimental basis have occurred, or can occur, and what must be done in the way of model experiments and reconstitution experiments.

I should also like to introduce sources of information other than experiments on the earth, namely, the exploration of the nearer celestial bodies—the moon, Mars, and Venus—to see what information we can hope to get from them in the near future that may bear on the overall question before us. And finally I should like to seek the humanistic implications of this exercise, if there are any.

REFERENCES

1. TINOCO, I., JR., HALPERN, A., and SIMPSON, W. T., The relation between conformation and light absorption in polypeptides and proteins. *Polyamino acids, polypeptides and proteins* (editor M. A. Stahlman), pp. 147–57. University of Wisconsin Press, Madison (1962).
2. BLOW, D., Unlocking the secrets of the enzyme. *New Scient.* **36**, 218 (1967).
3. DICKERSON, R. E., X-ray analysis and protein structure. *The proteins* (editor H. Neurath), 2nd edn., vol. 2, p. 603. Academic Press New York (1964).
4. BILTONEN, R., LUMRY, R., MADISON, V., and PARKER, H., Studies of the chymotrypsinogen family. IV. The conversion of chymotrypsinogen A to α-chymotrypsin. *Proc. natn. Acad. Sci. U.S.A.* **54**, 1412 (1965).
5. STERNLICHT, H., and WILSON, D., Magnetic resonance studies of macromolecules. I. Aromatic-methyl interactions, helical structure effects in lysozyme. *Biochemistry* **6**, 2881 (1967).
6. SCHEIBE, G., Wechselseitage Bindung und Energieübertragung in Molekeln in flüssiger phase. *Z. Elektrochem.* **52**, 283 (1948).
7. JELLEY, E. E., Molecular nematic and crystal states of 1,1'-diethyl-ψ-cyanide chlorice. *Nature, Lond.* **139**, 631 (1937).
8. WATSON, J. D., and CRICK, F. H. C., General implications of the structure of DNA. Ibid. **171**, 964 (1953).
9. CALVIN, M., and KODANI, M., The physico-chemical structure of the chromosomes. II. *Proc. natn. Acad. Sci. U.S.A.* **27**, 291 (1941).
10. —— From microstructure to macrostructure and function in the photosynthetic apparatus. *Brookhaven National Laboratory Symposia* **11**, 160 (1958).
11. —— *Chemical evolution* (Condon lectures). University of Oregon Press, Eugene (1961).
12. PETTIT, F. H., and REED, L. J., Keto acid dehydrogenase complexes. VIII. A comparison of dihydrolipoyl dehydrogenases from pyruvate and ketoglutarate dehydrogenase complexes of *E. coli*. *Proc. natn. Acad. Sci. U.S.A.* **58**, 1226 (1967).
13. FERNANDEZ-MORAN, H., VON BRUGGEN, E. F. J., and OHTSUKI, M., Macromolecular microscopy. *J. molec. Biol.* **16**, 141 (1966).
14. ASAKURA, S., EGUCHI, G., and IINO, T., *Salmonella* flagella: in vitro reconstruction and overall preparation of flagellar filaments. Ibid. **16**, 302 (1966).
15. WILLIAMS, ROBLEY C., Private communication from Virus Laboratory, University of California (1961).
16. MACLEAN, E. C., and HALL, C. E., Studies on bacteriophage φX–174 and its DNA by electron microscopy. *J. molec. Biol.* **4**, 173 (1962).
17. GOULIAN, M., KORNBERG, ARTHUR, and SINSHEIMER, ROBERT L., Enzymatic synthesis of DNA. XXIV. Synthesis of infectious phase φX–174 DNA. *Proc. natn. Acad. Sci. U.S.A.* **58**, 2321 (1967).

10

GENERATION OF MEMBRANE STRUCTURE

WE have reached the point at which the information-bearing type of structure links up with the catalytic structure. We can see that the structures had self-assembling qualities: that is, they were in such a specific form that once they were created in any open milieu and allowed to reach a high enough concentration, they would assemble themselves into structures of higher order. This self-assembly process took place not only among molecules of a single kind—proteins, nucleic acids, or polysaccharides—but could be extended to mixed assemblies of different kinds of polymers. This is only the beginning, and is only a part of the problem of how a cell construction could have appeared. Another essential step is required. This is the concentration of the material that is being generated, involving all the various processes outlined above, and the separation of that relatively concentrated collection of material from the open environment—from the sea itself, if you like—in order to give rise to a cell. The cell itself is, of course, subdivided into subunits or organelles. The membrane boundary of the cell itself is the one we shall discuss here, but we also know that inside the modern cell there are functional boundary membranes for almost all the cell organelles—the nucleus, the mitochondria, the chloroplasts, and so on.

Before exploring the physico-chemical processes that might operate on a complex mixture of macromolecules to give rise to such limiting membranes, I shall describe some of the enormous variety of membranes found in today's organisms in an attempt to discern any general principles of structure that there may be. To do this I shall not go through a catalogue of all possible membranes; I shall select a few that seem to represent typical membranes, together with those with which I am personally familiar. The first example of such a cellular membrane is shown in Fig. 10.1: the bacterial membranes from *Rhodopseudomonas spheroides*.

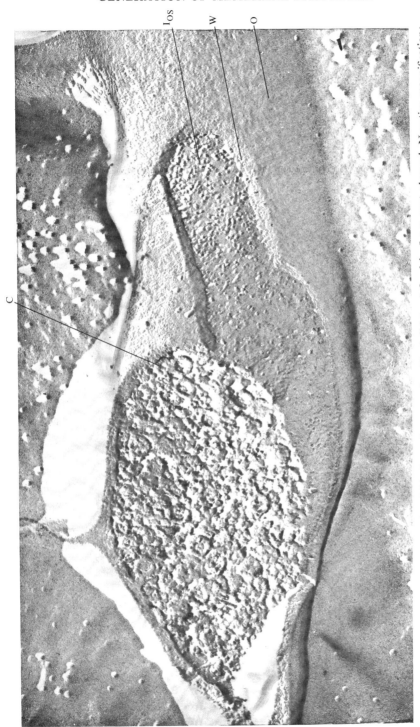

FIG. 10.1. Electron micrograph, freeze-etch, of *Rhodopseudomonas spheroides* (wild type) fixed in 5 per cent glycerol. Negative magnification: 30 times; photograph magnification: 100 times. *O*, outside surface of cell; *W*, outer wall in section; *I*os inner membrane, outer surface; *C*, internal structure (showing organized elements). Photograph by George Ruben, Laboratory of Chemical Biodynamics, University of California, Berkeley.

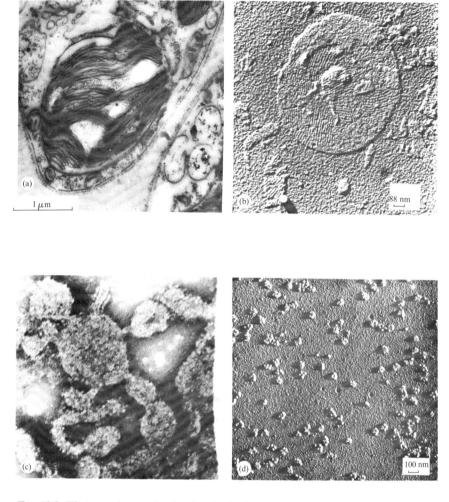

FIG. 10.2. Electron micrographs showing the fundamental particles of biology: ribosomes, electron transport particles of the mitochondria, quantasomes of the chloroplasts, and unit lipoprotein membrane. (a) Chloroplast with mitochondria *Chlamydomonas* (Sager); (b) quantasomes from spinach; (c) negatively-stained mitochondria (Park and Packer); (d) polysomes making haemoglobin.[1]

This freeze-etched photograph shows the several features of the membrane. The smooth area is the outside surface of the cell itself, which has been broken across the outer wall and the inner membrane. The cell wall and cell membrane are shown in some detail. The internal structure of the bacterium, which is not well shown in this figure, exhibits some chromatophore-like particles. Note the granular character of the cell-wall section as well as the particles distributed on the outer surface of the inner membrane.

Cellular organelles are many and varied, and Fig. 10.2 shows a collection of them—chloroplasts, quantasomes, mitochondria, and ribosomes.[1-5] The quantasome structure is the beginning of the really 'fine' structure of one of the lamellar layers of the chloroplast, with its ordered

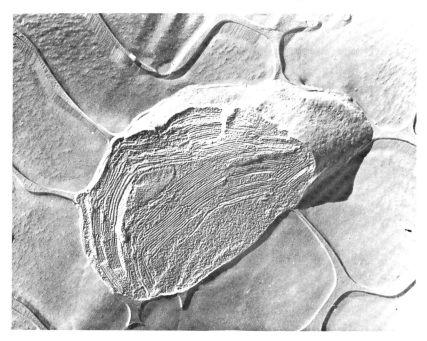

FIG. 10.3. Intact chloroplast from spinach; freeze-etched micrograph. Photograph by Kay Shumway, 1966. Magnification × 10 000.

array. In the negative-stained mitochondrial membranes an ordered linear structure also exists, and some of the particles are attached to the membrane itself. The ribosomes, shown in Fig. 10.2 (d), are the particles that assist in the synthesis of proteins.

I shall now focus a little more sharply on the chloroplast lamellae and membranes, because I think they represent in general principle the type of structure that most membranes have. Another reason for this choice is that the series of chloroplast pictures reproduced here is from our own laboratory. Fig. 10.3 is a freeze-etched micrograph of an intact chloroplast from spinach, which can carry out the full range of photosynthetic reactions at practically the same rate as the intact plant. It is easy to see the outer membrane of the chloroplast and also the inner lamellar structure, which is very highly developed. Several aspects of the lamellar membrane, as well as the limiting membrane of the chloroplast, are shown here and identified in Fig. 10.4.[6] The nature of the chloroplast lamellar membranes is shown in greater detail in Fig. 10.5,[7] which shows how they are put together. For many years these membranes were

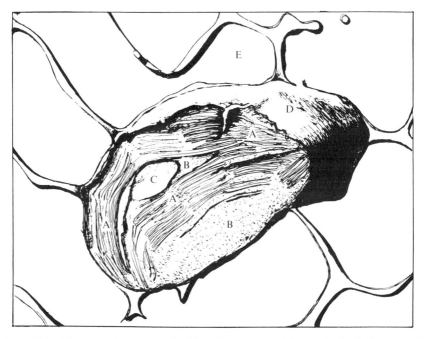

FIG. 10.4. Electron micrograph of chloroplast, prepared by method of Jensen and Bassham.[6]

   A, lamellae. Layers of lipid and protein with chlorophyll, other pigments, and enzymes. Site of light absorption and splitting of water to give oxygen and hydrogen.

   B, stroma. Soluble enzymes. Site of conversion of carbon dioxide and hydrogen to sugars and other products.

   C, starch granule. One of the storage products of photosynthesis.

   D, surface of chloroplast. The double membrane surrounding the chloroplast can be seen, with most of the outer membrane split off revealing the surface of the inner membrane.

   E, frozen solution in which chloroplast was suspended.

believed to be continuous layers of the lipid components, forming a monomolecular (really a bimolecular) layer on an interface; a phenomenon with which the physical chemists had long been familiar. As these membranes were examined more closely under the electron microscope, we began to see that while this particular kind of organization probably plays a role in the ultimate structure of the membrane, it is not the basic property that gives rise to the membrane. A closer view of one of the lamellar membranes, in Fig. 10.5, which shows the cross-

FIG. 10.5. Section of chloroplast.[7]

section of the chloroplast membrane, stained with uranyl acetate, reveals that the layers, roughly 10 nm (100 Å) apart, are not continuous, even in cross-section. One can see that the membranes are composed of subunits. This was already apparent in the chloroplast membrane shown in Fig. 10.2 (b), where the fine structure is visible when a lamella is viewed on its large surface. Fig. 10.6 shows the same type of thing very highly magnified, and one can see that it is highly and particularly structured. Even the particles that give rise to the main structural features are themselves made of some substructural entities. This is indicated in the circled unit, which is composed of at least four aggregated subunits. These small entities aggregate to intermediate sizes, which in turn aggregate to still larger ones, thus ultimately giving rise to the sheet-like structure.

Fig. 10.7 shows a freeze-etched picture of a group of chloroplast lamellae. The layers are evidently made up of at least two distinct sizes of particles.[8] The large ones are folded into the layer between two smaller ones; at each successive layer the double character can be observed. The chloroplast lamella must therefore be composed of at least two sizes of aggregating units, fitted together in a double aggregated layer. This is a rather complex situation; and Fig. 10.8 is a diagrammatic representation of a suggested structure for the chloroplast lamellae.[8] It is an aggregate of units containing various molecular materials. The chloroplast membrane is apparently an aggregation of one or several different subunits made up of protein coated with lipid, which have in them, probably on the surface, the porphyrin heads of the chlorophyll molecules. This diagrammatic representation of the chloroplast lamellar structure has been obtained by electron microscopy on the one hand and chemistry on the other. This particular structure seems to be a prototype: almost every other membrane that has been explored in sufficient depth to allow anyone to make a suggestion about its construction has given rise to this same type of structure—an aggregation of subunits. The same is true for mitochondrial membranes, erythrocyte membranes, neuronal membranes, and a wide variety of structures from the plant, animal, and microbial world.

Physical and chemical processes for concentration and separation by membranes

We do not have to formulate the membrane as, let us say, a continuum of phospholipid, but rather as an aggregation of identical subunits, each covered with the lipid membrane, which, upon fusion, gives rise to what appears to be a continuous bilipid membrane (a lipoprotein structure

PLATE 5

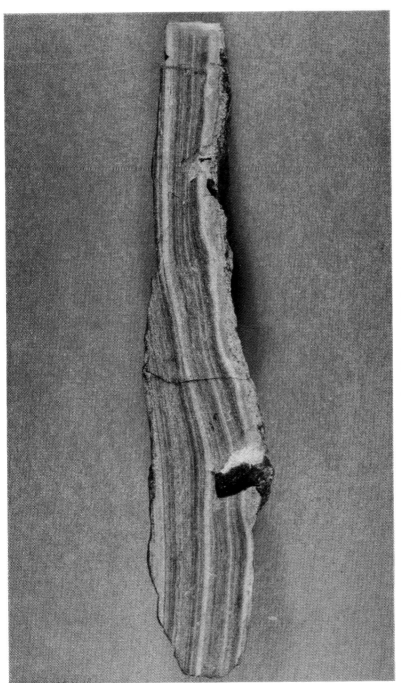

Green River Shale, polished section

PLATE 6

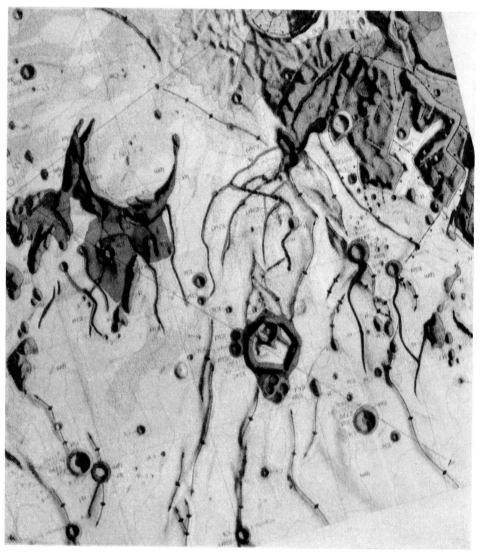

Map of part of the lunar surface corresponding to Lunar Orbiter IV, frame 51
(U.S. Geological Survey map, 1965)

PLATE 7

Photograph of a planet at 300 000 miles distance (NASA photograph)

PLATE 8

King Aroo cartoon by Jack Kent

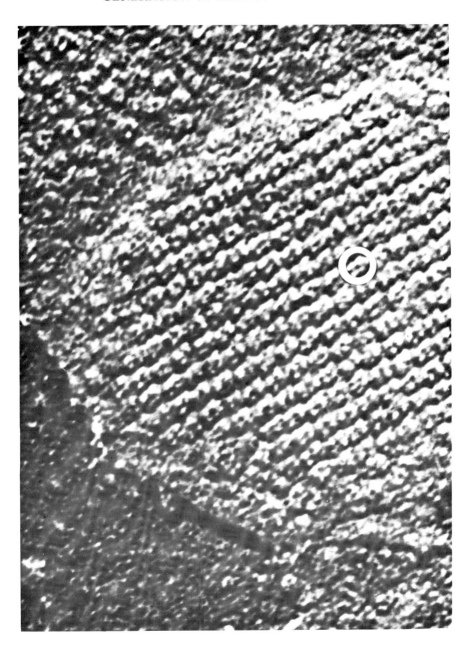

FIG. 10.6. Section of chloroplast membrane at high magnification. (Photograph by D. Healey.)

FIG. 10.7. Group of chloroplast lamellae, freeze-etched.

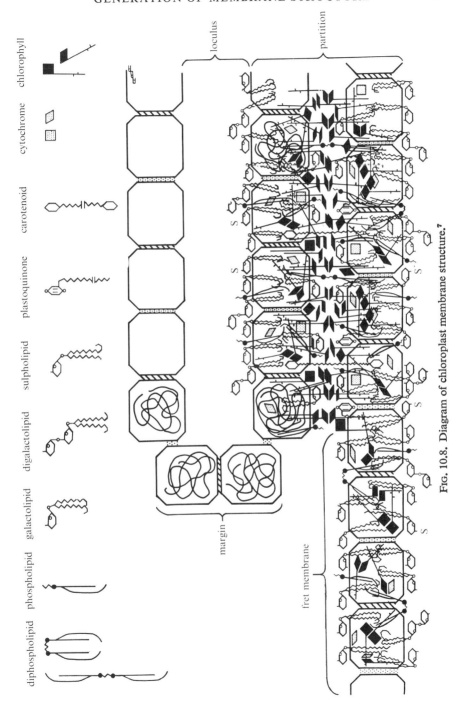

FIG. 10.8. Diagram of chloroplast membrane structure.[7]

made up of identical subunits). With this realization we can return to examine the physico-chemical processes that might operate on the polymeric collection of protein, nucleic acid, lipid, and polysaccharide to determine what kind of mixed events might be expected to occur.

At least four quite different, but not mutually exclusive, physico-chemical processes have been considered for the process of concentration, separation, and membrane development. These are evaporation, coacervation, adsorption, and film (or membrane) formation. Evaporation is the simplest concentration process. The suggestion made, and explored most extensively, by Fox is that the fairly dilute undifferentiated oceans, seas, or lakes would periodically be cut off in tidal pools and rocky shallows, and evaporation by solar or volcanic heat would gradually concentrate the polymeric materials to fairly high levels.[9] Fox performed experiments to show what would happen if a process of this kind actually occurred. He took mixtures of amino acids and heated them to 75–150 °C in the presence or absence of polyphosphate. He could then extract from the products a polypeptide mixture, which, upon being put into contact with water and cooling, would give rise to peculiar structures that he called 'microspheres'. These structures do not dissolve; they are roughly spherical structures with what appear to be surface boundaries. Fig. 10.9 shows an electron micrograph of some of these microspheres, which resulted from the heating of all 20 amino acids dissolved in excess aspartic and glutamic acids and allowing the solution to cool. Not only do they come out of solution as little droplets, but these droplets seem to have surface membranes. Fox has gone one step further with this type of experiment and has shown that some of these microsphere preparations have what he describes as catalytic capacities. For example, he measured the rate of hydrolysis of ATP under the influence of a suspension of such microspheres, which were made in the presence of zinc, and obtained what is obviously some kind of catalytic effect (see Fig. 10.10).[9] Clearly the microspheres must exercise some kind of catalytic function on the hydrolysis of pyrophosphate, but the question of whether the effect is specific or simply due to a local concentration, because of the excess solubility of the triphosphate in the proteinoid sphere, remains to be established.

The coacervate theory of separation has been mainly developed by the Russian biochemist A. I. Oparin.[10] Coacervates result when a solvent (for our purpose water) contains two different macromolecular polymers that interact with the solvent, but do not interact well with each other. Such a situation gives rise to a phase separation, and one phase will commonly

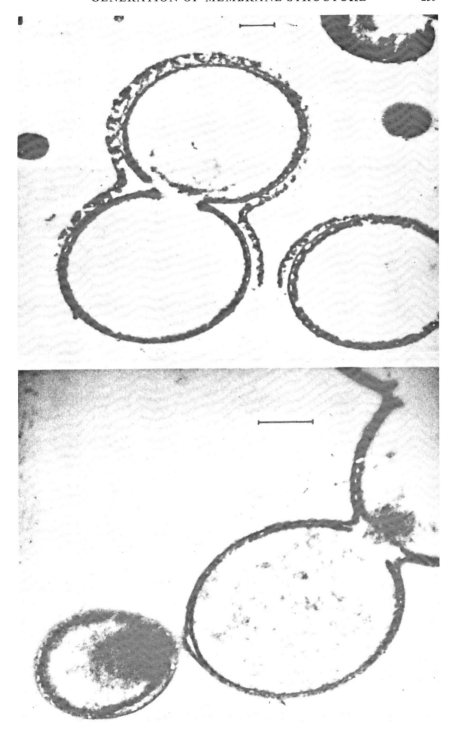

FIG. 10.9. Electron micrographs of double layers in microspheres.[9]

be a dispersed one suspended in the other as the continuous phase. One polymer is mainly in the dispersed phase and the other component is mainly in the continuous phase; in the cases that are of interest to us both phases are essentially water. In this instance such a coacervate has been made from polyethylene oxide and a polymeric sugar. The coacervation process gives rise to a phase boundary; small molecules in the

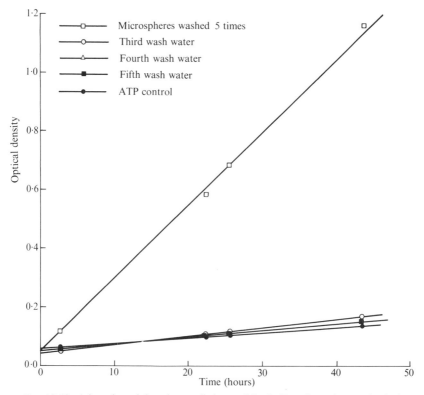

FIG. 10.10. Adenosine triphosphate-splitting activity in Zn-microspheres. Optical density is molybdate colour intensity, measuring release of phosphate.

aqueous system might distribute themselves unequally between the two kinds of aqueous phase.[11,12] This is essentially Oparin's idea of the beginning of a separation and delineation process. He points out that as the polypeptides and polynucleotides grow they become more liable to such a separation. He has done some interesting experiments in which he showed that if mononucleotides are polymerized in the presence of a basic polypeptide (either synthetic or natural), the solution, as the polynucleotide becomes larger, gradually acquires a milky appearance. Upon microscopic examination small coacervate droplets were seen in this milky-looking solution. The droplets could be separated out, and they were found to be made up primarily of the polynucleotide material. Oparin has followed up this idea and attempted to demonstrate that all sorts of enzymatic reactions may be facilitated by the presence of such coacervate systems.[10]

Adsorption is, of course, one of the other principal means of concentration and separation, primarily by adsorption on a solid surface from the liquid milieu of the ocean. This particular physico-chemical method is the one that has been favoured by J. D. Bernal for the last 30 years, and a variety of experiments have been performed, in England and elsewhere, in the attempt to demonstrate that such adsorptive systems (clays, for example) might lead not only to concentration but also to order.[13,14] (This same factor is also relevant in connection with a route to generating information-storing systems:[15] see p. 176.)

In general, the existence of a phase boundary, that is a boundary between two different physical phases (water phases as in coacervates, or any two phases), leads to the formation of a region of unique chemical composition and molecular structure between the two phases. This is something that occurs automatically the moment a system appears that has a phase boundary in it. It leads to the formation of a membrane, which is a universal characteristic of free-living organisms. Membranes not only separate living cells from the external environment, but also separate subenvironments within the cell.

The nature of membranes

The nature of the membrane is determined by the molecules available in the two phases—lipids, proteins, polysaccharides, polynucleotides—and also by the nature of the phase boundary at which the molecules can accumulate. For example, a liquid–liquid boundary, such as the coacervate boundary, gives rise to one kind of membrane structure, not different in principle from any of the others, but different in detail. The liquid–gas boundary (water–air) is found at the surface of the sea; the liquid–solid boundary is the one obtained upon adsorption on clays from the water layer; the solid–gas boundary was discussed above in connection with the influx of meteorites (p. 116).

Of these different phase boundaries, the most interesting from our point of view is the liquid–air interface. We have already explored in some detail the way in which surface films are formed in this kind of system. In a sense this is where we came in. At the beginning of the book we discussed the appearance of the long-chain hydrocarbons in a non-biological system. In the presence of CO_2 these long-chain hydrocarbons, which are formed readily in the primitive milieu, would have a polar end. They are members of a general class of compounds called *surfactants*; that is, they have a hydrophilic end (the carboxyl group) and a hydrophobic end (the hydrocarbon chain). This, in general, gives rise to the concentration of such molecules at phase boundaries, between water and air, or water and an organic solvent such as hexane. Fig. 10.11 shows

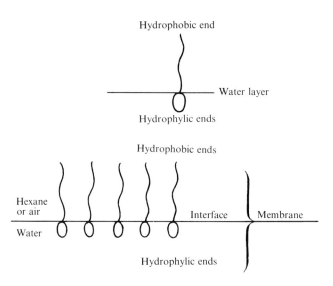

FIG. 10.11. Phase boundary marked by double-ended molecules with a hydrophilic and a hydrophobic end.

a generalized diagram of the phase-boundary principle operating on a simple bifunctional molecule. The hydrophobic end, the long hydrocarbon chain, sticks out of the water layer, and the hydrophilic end, containing the carboxyl group, is in the water layer. It was a principle of this kind that gave rise to the formulation of the bilipid membrane in the first place,[16] and it remains the basic idea for the development of the cell boundary.

These simple lipids can form monolayers. More complex lipids, such as the phospholipids, which have a somewhat larger polar chain, the end of which may actually be charged (negatively or positively), have also

shown some of the interesting 'constructional' capability for this type
of membrane. The presence of a variety of high polymers in the water
layer underneath the interface (already containing the hydrophobic and
hydrophilic groups) will affect the way in which the membranes are
formed, and the way in which they collapse. If we imagine that such a
phase boundary is formed and then disturbed, for example by wind

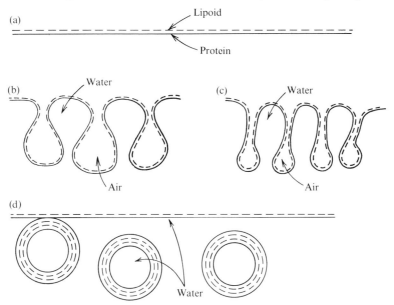

FIG. 10.12. Droplet formation by film collapse.

blowing along the surface, we can see how the membranes might collapse and form closed vesicles. This phenomenon happens more readily
and in a more defined way if other kinds of macromolecules are present
(for example protein in the water layer) that can keep the film from
collapsing upon itself. Moreover, the films will be unsymmetrical because
one side will be hydrophobic and the other hydrophilic. At the water–
air interface, which is mechanically disturbed, the membranes will collapse. Fig. 10.12 shows droplet formation by film collapse.[17] The vesicular
structures contain either water or air, depending upon the way in which
they are compressed and the relative amounts of lipid and protein that
are associated in the interface boundaries. The nature of the vesicle
boundaries will depend on the particular molecules involved. Some of
these vesicles have been constructed in the laboratory in recent years. At
Babraham, near Cambridge, there has been an interesting series of
studies involving the reconstitution of limiting membranes with phospho-

lipids and with suitable proteins such as cytochrome c. The mechanical disturbance of the lipoprotein layer gives rise to closed vesicular structures of various sizes, shapes, and dimensions. The result of this type of experiment is shown in Fig. 10.13.[18]

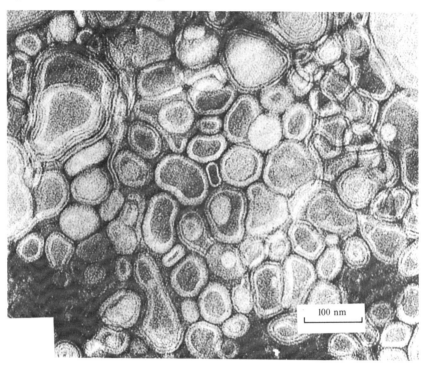

Fig. 10.13. Experimental formation of closed vesicular structures in lipoprotein layer. Phosphatidyl choline–phosphatidylethanolamine (1 : 1) mechanically shaken in presence of cytochrome c (0.03 μmol per μmol of lipid), dispersed in ammonium acetate and subsequently stained with ammonium molybdate. Reference bar is 0.1 μm long.

It is clear that the lipid boundaries of some vesicles can be fairly thick. The simplest vesicle in Fig. 10.13 is about 30 nm (300 Å) from one end to the other in the centre; so the boundary layer of the vesicle is about 6–8 nm (60–80 Å) thick: just the thickness of most lipid layers in nature. These layers are very interesting; not only do they have the particular vesicular structure we are seeking, bounded by a lipoprotein membrane, but they are also osmotically active and selectively permeable, and are thus beginning to show many of the properties we should expect to find, and do find, in natural membranes.[19,20] Here they are being formed from the froth of the ocean, so to speak, by the wind.

This is really about as far as we can go with models of this kind of construction work. A number of reconstitution experiments have, however, been tried with natural membranes, like the ones that have been so successful in the reassembly of subunits of enzymes and of the subunits of viruses (see pp. 210, 216). Self-assembly even to the unsymmetrical membrane level, then, is a factor that can be built into the structure of the molecules. We have gone so far in the membrane story with purely synthetic materials; and there is obviously much more model work to be done. The dismantling and reassembling of biologically important and active membranes has only just begun. The most extensive work of this kind so far published is that of David Green at Wisconsin, who believes he has dismantled and reassembled the whole oxidative system in mitochondria.[21] He also thinks he has dismantled and reassembled the corresponding membranes of the chloroplast. We have not yet succeeded in doing this ourselves, but I think it is only a matter of time and technique until this kind of thing can actually be done in the laboratory. When the various parts of such things as mitochondria and chloroplasts can be taken apart and reassembled in a functional way, we shall then know more clearly how these things really perform their multiple tasks.

It is worth re-emphasizing the fact that the subunit structure will determine not only the geometry of the membrane but also its function. The reassembly process of external membranes has also been begun. Fig. 10.14 shows the reassembly of the outside limiting membranes of the smallest free-living organism that we know so far, the mycoplasma.[22] Fig. 10.14 (a) shows the initial organism, with just the cell wall; Fig. 10.14 (b) shows the isolated membranes. These are shown reconstructed in the lower picture after having been dismantled by the addition of excess detergent; to reassemble the membrane the detergent was diluted out. There is really not a very spectacular resemblance between the reassembled membrane and the initial one, but it is a move in the right direction. There have been other experiments of the same kind in which the reassembled membrane has been created by allowing the assembly to occur at a phase boundary. When that happens, the permeability characteristics, electrical properties, and so on, of the membranes can be measured. Some progress has been made in this direction.

Summary of chemical evolutionary processes

We have tried to carry out an exercise in the application of our general knowledge of physics and chemistry and the structural and functional properties of the simplest of biological systems, to see if we could reconstruct at least one possible route from the primeval molecules on the

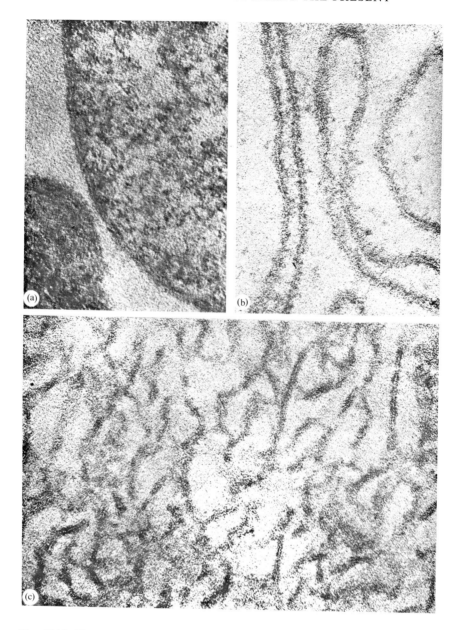

FIG. 10.14. Electron micrographs of thin sections of (a) cells of *Mycoplasma laidlawii*, (b) isolated membranes, (c) reaggregated membrane material. Magnification ×113 000.[22]

surface of the earth to something that one might accept as a living organism. We started with the primeval molecules, and by putting in energy have created the monomers: metabolites and energy-storage molecules, such as the carbon–nitrogen multiple bond in cyanide and the pyrophosphate bond in ATP. With additional energy, either in the form of light or from energy-storing molecules (cyanide and pyrophosphate), we have polymerized these monomers (amino acids, sugars, nucleotides, fatty acids) to the corresponding polymers (nucleic acids, proteins, lipids, polysaccharides). Following this polymerization, or simultaneously with it, autocatalysis came into play, resulting in a selection process that gave rise to two streams of polymers. One of these is a poorly reproducing but catalytically effective system (the enzyme proteins); the other a very accurately reproducing but poorly catalytic system (the nucleic acid information-storage systems). We then devised a means of coupling these two systems together so that the fidelity of information transfer and the facility of catalysis and energy transduction could both survive. This latter process gave rise to what is essentially a virus particle. This coupled information and enzymatic system, under the influence of phase-boundary separations, could ultimately give rise to a cellular structure, encased in a boundary membrane.

This is certainly a conceivable sequence. I see no really fundamental mathematical or philosophical argument that could deny it. However, whether it is the *only* possible sequence, given the nature of the elements and the physical environment, or even whether any such sequence is a necessary event, given the initial conditions is, of course, the problem. But such a question provides part of the stimulus for our lunar and planetary exploration systems, from which we might hope for an answer.

As long as we are limited to biology as it is on the earth, it is going to be very difficult for us to be sure that such a system occurred in the way described in this book.[23] We shall have to find other places in the universe, preferably near by in our solar system, in which this process is going on and has not gone all the way, so that we can see it at some other stage of its development. This is one of the reasons why I, at least, am interested in the lunar and planetary exploration programme: I think that answers to this fundamental question may be provided by such exploration.

Lunar and planetary exploration

Some small items of information have already come from the experiments that have so far been done, and the imminent exploration of the moon may provide some additional important clues, if not complete answers. One of the reasons why I am so optimistic about this possibility is shown in one or two illustrations. Pl. 6 shows a map of a certain region of the moon that was made from a telescope in 1965.[24] The half-complete crater toward the lower right and the totally completed crater near the centre, with another one inside of it, are characteristic lunar

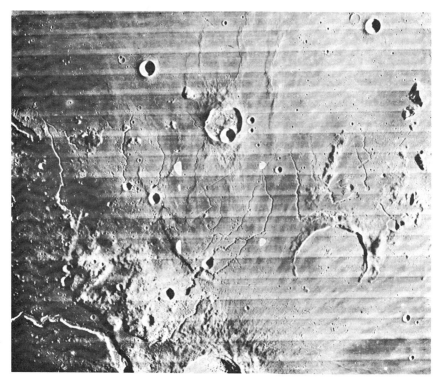

FIG. 10.15. *Lunar Orbiter IV*, frame 51. Aristarchus crater, showing channels and cuts (NASA photograph).

structures. This map was made from telescopic observation only, and the question now is, what is the surface of the moon really like? This is shown in Fig. 10.15. This was taken by *Lunar Orbiter IV* (frame 51) and indicates the channels and wall cuts; one can see the identical craters that were observable in Pl. 6. Some pecularities are visible here. There is a hole in the left side of the centre crater, which looks as though it had been worn down, and coming from the crater there is a rill that ends in the lunar desert. It looks as though a hole has been worn in the side of the crater and something has run out and gradually disappeared into the desert. When this was first observed it seemed to recall water over-flowing the walls of the crater. The fact that there might have been water on the moon at one time is the important conclusion that Harold Urey drew from this photograph.[25] If there was water anywhere on the moon, obviously there was water everywhere on the moon; and if there was water ever on the moon, it is still there, but it must be from 1 to 9 metres below the surface, in the form of an ice table. Since the temperature drops to -30 °C within some 0·5 m of the lunar surface, one can calculate diffusion rates and come to the conclusion that there will be a residual ice table, certainly at a depth of 9 m and perhaps only 1 m from the surface. If there is such an ice table on the moon there will undoubtedly be organic matter on it, even if it is 'only' the organic matter that the moon has swept out in the course of its orbital history, in the form of, say, carbonaceous chondrites.[26] However, this organic matter would have been preserved in its original form, in cold storage, so that we can see what it is.[27-30] We could then see something of this evolutionary history in the moon, and this is one of the reasons why we hope to get back some lunar samples and look closely at them.

How are we going to obtain these lunar samples? A digging machine has already been sent to the moon, and holes have been dug in the lunar surface. However, another method was chosen to bring back the lunar samples: to send a man. We have been heavily engaged in getting ready for this event, which should take place in the next year or two. Fig. 10.16 shows an astronaut training on a simulated lunar landscape in Arizona, dressed in an inflated space suit, with the instrument package and a sample box to contain the lunar rocks. The details of the sample box are shown in Fig. 10.17; the box was not easy to design, because of the difficulties of opening and closing it on the lunar surface, and because of the difficulties of manœuvring in the space suit, for example. We believe that the astronaut will return to the earth with about 35 (earth) kilogrammes of lunar rocks.[31]

Of course, it would be much better if we had some direct communication from elsewhere in the universe telling us about this evolutionary history. Recently some curiously pulsed radio signals have been reported, which are hard to interpret. These radio signals are in the wavelength region from 8–120 MHz, and they arrive with a periodicity of 1·33729 s. The duration of each pulse, as it comes in, is about 0·3 s; as it is broadcast, it is 0·016 s. (The Doppler broadening makes it appear to last 0·3 s after it reaches the earth.) The pattern of radio signals is of varying intensity, and the pulse train lasts for about 1 min before the intensity is

FIG. 10.16. Astronaut in suit, with lunar tool-chest (NASA photograph).

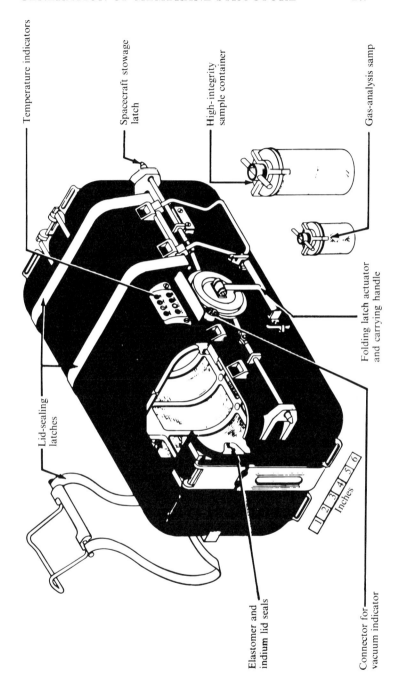

FIG. 10.17. Details of sample box for lunar rock collection (NASA photograph).

reduced to the level of the noise in the radio telescopes presently used.[32,33] It remains like that for several minutes, and then the intensity rises again, and the whole sequence goes on. Since the initial report there have ensued a whole series of explanations of one or another characteristic of the signals. I suspect that many people are busy trying to decode any information there might be in these signals, and are certainly trying to understand their origin.

I should like to end this chapter by referring to another experiment. A vehicle sent off recently from Cape Kennedy approached to within 300 000 miles of a planet and took a photograph, which is shown in Pl. 7, and from that planet a message was recorded. That message is as follows:[33]

```
111100001010010000110010000000010000010100
100000110010110011110000011000011101000000
001000001000010000100010101000010000000000
000000000010001000000000010110000000000000
000000010001110110101101010000000000000000
000010010000111010101010000000000101010101
000000000111010101011101011000000001000000
000000000001000000000000001000100111111000
001110100000101100000111000000001000000000
100000000100000001111100000010110001011100
100000001100101111101011111000100111111001
000000000001111100000010110001111111100000
100000110000011000010000110000000011000101
001000111100101111
```

There is a total of 551 0s and 1s. What does it tell us?

REFERENCES

1. CALVIN, MELVIN, The path of carbon in photosynthesis. *Science* **135**, 897 (1962).
2. SAGER, R., and PALADE, G. E., Structure and development of the chloroplast in *Chlamydomonas*. I. The normal g̣en cell. *J. biophys. biochem. Cytol.* **3**, 463 (1957).
3. PARK, R. B., and PON, N. G., Correlation of structure with function in *Spinacea oleracea* chloroplasts. *J. molec. Biol.* **3**, 1 (1961).
4. —— and PACKER, L. E., Unpublished results.
5. WARNER, J. R., RICH, A., and HALL, C. E., Electron microscope studies of ribosomal clusters synthesizing hemoglobin. *Science* **138**, 1399 (1962).
6. JENSEN, R. G., and BASSHAM, J. A., Photosynthesis by isolated chloroplasts. *Proc. natn. Acad. Sci. U.S.A.* **56**, 1095 (1966).
7. WEIER, T. E., and BENSON, A. A., The molecular organization of chloroplast membranes. *Am. J. Bot.* **54**, 389 (1967).

8. BRANTON, D., and PARK, R. B., Subunits in chloroplast lamella. *J. Ultrastruct. Res.* **19**, 283 (1967).
9. FOX, S. W., Simulated natural experiments in spontaneous organization of morphological units from proteinoid. *The origins of prebiological systems and of their molecular matrices* (editor S. W. Fox), pp. 361–73). Academic Press, New York (1965).
10. OPARIN, A. I., The pathways of primary development of metabolism and an artificial model of this development. *The origins of prebiological systems and of their molecular matrices* (editor S. W. Fox), p. 331. Academic Press, New York (1965).
11. SMITH, A. E., BELLWARE, B. T., and SILVER, J. J., Formation of nucleic acid coacervates by dehydration and rehydration. *Nature, Lond.* **214**, 1038 (1967).
12. LIEBL, V., and LIEBLOVA, J., Coacervate systems and life. *J. Br. interplanet. Soc.* **21**, 312 (1968).
13. BERNAL, J. D., *The physical basis of life.* Routledge & Kegan Paul, London (1951).
14. —— *The origin of life.* World Publishing Co., New York (1967).
15. CAIRNS-SMITH, A. G., The origin of life and the nature of the primitive gene. *J. theoret Biol.* **10**, 53 (1966).
16. DAVSON, H., and DANIELLI, J. F., *The permeability of natural membranes.* Macmillan, New York (1943).
17. GOLDACRE, R. J., Surface films, their collapse on compression, the shapes and sizes of cells and the origin of life. *Surface phenomena in chemistry and biology*, pp. 278–98. Pergamon Press, London (1958).
18. PAPAHADJOPOULOS, D., and MILLER, N., Pholid model membranes. I. Structural characteristics of hydrated liquid crystals. *Biochim. biophys. Acta* **135**, 624 (1967).
19. BANGHAM, A. D., DE GIER, J., and GREVILLE, G. D., Osmotic properties and water permeability of phospholipid liquid crystals. *Chem. Phys. Lipids* **1**, 225 (1967).
20. PAPAHADJOPOULOS, D., and WATKINS, J. C., Phospholipid model membranes. II. Permeability properties of hydrated liquid crystals. *Biochim. biophys. Acta* **135**, 639 (1967).
21. GREEN, D. E., ALLMANN, D. W., BACHMANN, E., BAUM, H., KOPACZYK, K., KORMAN, E. F., LIPTON, S., McCONNALL, D. G., McLENNAN, D. H., PERDUE, J. F., RIESKE, J. S., and TZAGOLOFF, A., Formation of membranes by repeating units. *Archs Biochem. Biophys.* **119**, 312 (1967).
22. RAZIN, S., MOROWITZ, H. J., and TERRY, T. M., Membrane subunits of *Mycoplasma laidlawii* and their assembly into membrane-like structures. *Proc. natn. Acad. Sci. U.S.A.* **54**, 219 (1965).
23. For a discussion of life based on elements other than carbon, see WALD, GEORGE, The origin of life. Ibid. **52**, 595 (1964).
24. MOORE, H. J., Geologic map of Aristarchus region of the Moon. *Map I-465* (*LAC*-39), U.S. Geological Survey (1965).
25. UREY, H. C., Water on the Moon. *Nature, Lond.* **216**, 1094 (1967).
26. VAUGHN, S. K., and CALVIN, MELVIN, Extraterrestrial life: some organic constituents of meteorites and their significance for possible extraterrestrial biological evolution. *Space research* (editor H. Kallmann), vol. 1, p. 1171. North-Holland, Amsterdam (1961).
27. SAGAN, C., Indigenous organic matter on the Moon. *Proc. natn. Acad. Sci. U.S.A.* **46**, 393 (1960).
28. SAGAN, C., Origin and planetary distribution of life. *Radiat. Res.* **15**, 174 (1961).

29. ANDERS, E., The Moon as a collector of biological material. *Science* **133**, 1115 (1961).
30. LEDERBERG, J., and COWIE, D. B., Moondust. ibid. **127**, 1473 (1958).
31. MCLAINE, J., Collecting and processing samples of the Moon. *Astronautics and Aeronautics*, Aug. 1967, p. 34.
32. HEWISH, A., BELL, S. J., PILKINGTON, J. D. H., SCOTT, P. F., and COLLINS, R. A., Observation of a rapidly pulsating radio source. *Nature, Lond.* **217**, 709 (1968); LAYZER, DAVID. The nature of pulsars. *Nature, Lond.* **220**, 247 (1968).
33. DRAKE, FRANK, Project Ozma. *Sky Telesc.* **19**, No. 3 (1960); *Physics today* **14**, No. 4, 40 (1961); TOVMASYAN, G. M. (ed.), *Extraterrestrial civilizations.* (Proceedings of First All-Union Conference on Extraterrestrial Civilizations and Interstellar Communication, Byurakan, May 1964). Israel Program for Scientific Translations, Jerusalem (1967).

PART III

THE VIEW FROM THE PRESENT
TOWARDS THE FUTURE

11

THE SEARCH FOR SIGNIFICANCE

WITH this brief exploration into an area of conjecture based on knowledge of physico-chemical facts, I conclude this part of the search for the interface between the non-living and the self-perpetuating living. What is the reason for this pursuit, and where can it lead? Perhaps it can be best summarized by man's need to search for significances, in his own life, and in life itself.[1]

Man's search for significance in living begins in the cradle, and sometimes reaches fever-pitch in early adulthood. By the time the child has grown to the man there are as many answers to these questions about the significant as there are persons who seek, and for each person there are sometimes as many answers as there are hours to compound them.

Clearly all human beings are engaged from birth in reaching conclusions about their experience in relation to their environment. For the most part, the criteria used in this process are acquired so gradually and so subconsciously, from the environment in which we grow, that it is possible for a man to live out his span of years without ever feeling the need to examine rigorously—perhaps to challenge and ultimately to accept, revise, or reject—those guidelines by which he has been assuming that he could safely and successfully conduct his life.

The chief business of any living thing is survival, as an individual and as a species. The survival of progeny can properly be considered part of the survival of the individual, for part of the individual lives on in genetic material and in transmitted knowledge in each new individual, whether microbe or man.

Many a man, once survival is reasonably assured, gives up his own quest and accepts the external standards that have become generally accepted currency for measuring success in competition against other men who are also striving to survive. This success may be measured in 'things'—houses, cars, sheep—or in the less tangible and sometimes more fickle approval of a man's peers or the larger public. This differs little from the standards of the wolf pack, where the strongest eats first from the kill, and the others follow, each according to his strength. But man has, perhaps, a business beyond that.

We who are born with the juxtaposed thumb; the slender, upright, interlocking spine; the bone-encased computer for storing, sorting, and weighing every sensory impression, every idea engendered by these impressions, and every fantasy unrelated to consciously observed things; we must wonder how we came to be; where we fit with other men; what are the rights and duties which constrain our acts; and what is the relationship of this unique aggregate of atoms (known by our name) to the rest of the universe. We must wonder what is the nature of truth and what is its duration. We must wonder at the nature of the God who was once the small property of a small tribe, and now orders a cosmos.

The thoughtful man seeks a reason to live out his days, a reason and a method for mankind to survive. He seeks a relationship between mankind and the cosmos, which he still only dimly perceives, for man can dream beyond his capacity to produce and beyond his ability to understand consciously. Man has seen the stars, and is therefore a pursuer of the cosmic dimension too. Each new crumb of knowledge holds no firm promise to answer his questions, but offers only another door on a route on which new doors are created by the facts that opened the last one.

For man to search for those values that are beyond his own survival he must, of necessity, establish a relationship with the thoughts of other men. There is no sure path for any man, be he scientist, artist, or artisan, but each must know the state of his own art and be the master of his own conscious self, and he must possess a sense of humility that can open his mind to new insights. The true student will seek evidence to establish fact rather than to confirm his own concept of truth, for truth exists, whether it be discovered or not. And the wise student is aware that truth can sometimes wear a different robe in the eye of each beholder. The prospects for a revelation of ultimate truths are nebulous indeed.

To our universe we are related in scale in the same measure as microbe to man, electron to microbe. But the size relationship is negligible. As men we seek a reason for living beyond mere survival: a rationale to justify our apparent dominance over all other forms of life. We seek a relationship between man and man, and between man and a cosmos whose limits we cannot comprehend.

For a scientist, it is often difficult to recognize the significant when it appears, even in his own area of competence. For the man whose life is spent with science, it is at least as difficult as for other men to set a value-judgement on some other phase of life. To examine my own search for significance, I shall first try to interpret the meaning of this search in my own terms. Secondly, I shall attempt to look into the experience of a scientist, and of science, to determine whether there is anything in the content or method of science that might help us to establish significant goals in the general sense. Finally, I should like to examine in retrospect what seems to me to have been the 'guiding star' of my own life.

A god whom men conceived in man's own image, and whom we confined and imprisoned in our small world, was both the foundation and the star of the Western world for the last 2000 years. This positive image provided an unambiguous guide for man's best activities, whether he were farmer, monk, artist, soldier, or king. For almost 2000 years there was relatively little ambiguity about what the mass of people should choose as the guiding star in their lives. For example, one need only recall the enormous number of men who participated in the Crusades, the chief subject-matter of the medieval and Renaissance artists, the building of cathedrals at the cost of comfort and well-being to all. All these activities had to do with the service and worship of a god. Even the scientist or natural philosopher of that period did his work in the name of his god.

Today, no such unambiguous star rides the heavens to direct our steps, either individually or collectively. Man's very concentration upon the need to search for significance, the broad growth of the existentialist philosophy over the last 20 or 30 years, and national and world-wide discontent and anxiety—all these things are evidence for this.

The very difficulty of finding an unambiguous guide for our social and national behaviour is evidence of problems of this same kind. We no longer rely with security on the old guides, and at the same time we face new problems, in part brought about by science and technology.

Science and technology

Science, for example, over the past few hundred years has been responsible for major changes in our view of ourselves. I should like to quote from the *Carnegie Review* (no. 2, 1964/5):

> With Copernicus and Galileo, man ceased to be the species located at the center of the universe, attended by sun and stars.
>
> With Darwin (and Pasteur, in spite of himself) he ceased to be the species specially created and specially endowed by God with soul and reason.
>
> With Freud he ceased to be the species whose behavior was governed by conscious, rational mind. . . .
>
> And, as we begin to understand and produce mechanisms that think and learn, man has ceased to be a species uniquely capable of complex, intelligent manipulation of his environment.

The age of Big Brother, described by Orwell in *1984*, is all too evident, even today. Concepts introduced by science have changed man's relationship to the universe from central to peripheral, at least in terms of matter and energy. This book is concerned with the concept that life itself is, or may be, a result of the behaviour of atoms and the architecture of molecules. These last two concepts almost unaided place man in a new relationship to his god and challenge the notion of his supremacy in the universe.

What has technology added to this confusion and problem? It has added entirely new dimensions of problems, with which we have to deal at the same time that the guidelines have been made less clear. Improvements in health and agriculture have not only lengthened the individual's life-span but have also brought the pressure and the impersonal tyranny of numbers. This impersonal tyranny makes it impossible to blame an individual for one's troubles—one can blame only a diffuse population. Improved communications have not only opened the world to the individual, but have also opened the individual to the world, as well as introducing the additional frictions of contact.

With all the diffuseness of our guides and with the introduction of new and complex problems, the problem of methods and direction becomes urgent. Is it possible that science, and its child, technology, can provide new or unique help in seeking alternative goals?

In order to learn whether alternative goals can be derived from science and technology, it is necessary to explore the nature of values and of significance. We must ask this question both in terms of personal values and in terms of social values.

There are, perhaps, two arbitrary divisions that might simplify the discussion. In one division the fact, or the idea, is judged by the individual in relation to his personal experience with his personal world. In the second, the judgement is made either by an individual or by a group in relation to the experience of mankind as a whole. Both kinds, and the gradients between and overlapping, are required of mature and intelligent men.

In retrospect, all of us can sometimes look back in our lives and identify those incidents that defined our course. We can follow the cause-and-effect relationships that gradually conditioned each new choice, and we view these choices as significant and endow them with meaning that makes them into symbols. For an infant the act of crying (initially a simple response to pain or discomfort) is quickly equated with meaning and importance when he finds that his pain or discomfort are usually relieved by the appearance of his mother in response to his cry. As he matures, other symbols acquire significance in relation to his own personal experience: the church, the flag, the number on the calendar that marks his birthday, the yellow envelope of the telegram, the books on a shelf, a man in a white coat: these are all symbols with meaning. In due course these symbols are re-evaluated and replaced in conscious importance, at the college level, by the score in the entrance examination, the draft card, the days on the calendar that designate the semester break, blonde hair, long hair and sandals, the Democratic Party, pot, Camus, or the bishop. These, too, will pass.

What is it, then, that describes this search, by which the trivial decisions—and the enduring ones also—are made? What can a scientist bring to this long search? From my limited experience I would offer two points of view: first, the premiss that man must always search, and that, to embark on a search, he must be prepared. Second, the future is made clearer only by familiarity with the past. From the scientific past as I know it, I should like to explore a few concepts that I think may have some significance for the future of mankind.

The very nature of matter as we know it today presents to us a universe in constant motion. The particles that make up my person are never static, whether they be viewed as individual atoms or as an integrated aggregate. The identity that is 'me' is only statistically the same as that which existed a moment ago, and the laws of probability make it most unlikely that the precise relationship of atoms that makes up my person will ever again be duplicated.

Although we describe the orbits of the planets, we cannot place them in infinite space, nor can we know that the processes that formed them are not at this moment proceeding toward their eventual disintegration. So it is not surprising that man, himself in flux, always asks the next question almost as he conceives the answer to the current one. But each question builds upon the answers of yesterday, and if truth is found to be clothed in a different robe each day, how do we search?

For the search that fills our years we need special equipment: the ability to record observations carefully and with great integrity; the power to evaluate new facts and compare them with what already is known; the courage to explore, and the willingness to make mistakes.

I should like to give some idea of my own procedure in this exploration, not so much in tactics, but in strategic planning. A good introduction to this is the comic strip about King Aroo illustrated in Pl. 8. The original drawings are in my office, a personal gift from King Aroo's creator, Jack Kent. The conversation is conducted between one Professor Yorgle and King Aroo. Jack Kent has caught the unpredictability and interrelatedness of all scientific search, a specific example of which is the way I came to be absorbed in the study of chemical evolution. Some years ago I became involved in questions of radiation chemistry, involving the effect of high-energy radiation (such as ultraviolet light, ionizing radiation from radioactive particles, or ionizing radiation from a particle accelerator) on molecules. How do the molecules behave when so much energy is poured into them? I was interested in the fundamental question of the interaction of radiation with the molecules, and I wanted to try to discover something of the way in which an electron, loosened by this irradiation, might chase its way up and down a molecule; this was my specific technical interest.

About that time someone had the good sense to introduce me to George Gaylord Simpson's lectures *The Meaning of Evolution*. I read this book, which was my first intimate contact with the question, written in a beautiful style, the most lucid exposition of the subject that I have encountered. The question on which my mind focused was not the sequence of the evolution of species but something further back. In his first chapter Simpson discusses the origin of life itself. I was already prepared for this question, because for a dozen years or so I had been trying to learn how green plants work and I was beginning to feel that this was a complex problem. Even though all animal life on the earth today is dependent upon green plants for its sustenance, it did not seem to me that green plants could have preceded animals—they were too complicated. The whole question, then, of which came first, plant or animal, interested me, and I was asking this question in a very naïve way. Simpson focused my ideas sharply on life itself. That, taken together with radiation chemistry, helped me to develop questions of different kinds, such as: 'did radiation chemistry produce the molecules necessary for the living organism?' This, in turn, gave rise to a whole chain of inquiries which are still going on today in the laboratory.

Another line of events converges on the same question. In 1957 a new object, *Sputnik*, was added to the orbit of the earth, which strangely affected man's view of what he could do. Immediately the question of the origin of life on the earth became much more concrete: it became subject to some kind of experimental test. If we can go to other places in the universe besides the earth, perhaps we can also find out something about how life got started on earth. The whole pattern of thinking focused then on to one thing.

The various pathways, quite diverse in their origin, gradually evolved into a pattern of work that is now not just localized in my own hand but active in many other laboratories throughout the world in experimental and practical ways. I give this as a practical example of how going in the wrong direction can bring one to somewhere else, as Professor Yorgle says.

I have just discussed one of the 'strategies' of science. There is another value that lies in the search itself: it is personally rewarding. There is, of course, a personal reward in the continuing search for truth as there is in the pursuit of any art, and it is a common observation that the able man becomes better as he pursues his art. There is some interesting physical evidence of the change that takes place in the brain during the process of learning. For example, just as muscles grow stronger when they are exercised, so there is now very definite physiological evidence showing that if a rat lives in a stressful environment, in which he has to use all his ingenuity to survive, his brain matures as a more functional instrument than the brain of a rat born and reared up in a completely undemanding environment.

Society also benefits from the search in its physical welfare, which we are often tempted to consider as the only goal of science and technology, and the only successful basis on which to seek college funds from alumni or governments. It is an observation that we accept daily as premiss for foreign aid to underdeveloped nations, for compensatory education programmes, for early training in foreign languages, and so on.

The brain has capacities for usefulness that are scarcely tapped by the challenge of simply staying minimally alive. A culture based on early exercise of mental ability achieves a high level of personal and cultural well-being, on which each generation continues to build.

When Faraday discovered electromagnetic induction, which was a strictly independent observation, no one could have imagined that this phenomenon would ultimately lead to the congestion of the modern city. When Pasteur discovered microbes, one might have conceived that this would ultimately lead to the alleviation of diseases—and that rather quickly—but that it might lead to planned parenthood was not so obvious. Finally, when Morgan rediscovered Mendelian quantized heredity, it was a far cry from today's stockpiled grain and the famines of the East.

Each one of these independent discoveries has given rise to quite effective and profound changes in our environment, and we can expect this process to continue.

Personal experience

I now come to the last element in this structure. Following this attempt to draw from the scientific experience some generalized concepts, it seems worth while to relate some personal history.

The fundamental conviction that the universe is ordered is the first and strongest tenet. As I try to discern the origin of that conviction, I seem to find it in a basic notion discovered 2000 or 3000 years ago, and enunciated first in the Western world by the ancient Hebrews: namely, that the universe is governed by a single God, and is not the product of the whims of many gods, each governing his own province according to his own laws. This monotheistic view seems to be the historical foundation for modern science.

A second tenet seems to lie in a fundamental quality of the human mind, which has been built into it by natural selection over the millennia of its evolution, namely, the need to know and understand. How this need to know and understand was evolved is probably traceable to some survival-selection pressure in the primitive being that gave rise to mankind. Be that as it may, it is here today and has been the source of man's greatest achievements in all areas of his activity: religion, the arts, the sciences, and the like.

When this 'need to know' is coupled with the conviction that it is worth knowing—that the search, in principle at least, is leading towards an understanding of an unattainable infinity of knowledge that is the law of the universe—a code of conduct for the hour, the day, the week, the year, the life comes forth.

When one is discouraged with the mass of scientific or sociological data and the need to encompass and comprehend all that man knows already, it is time to sit back and meditate on the simple one-celled amoeba. The amoeba has only one cell, but it can make itself x times over in an hour; it can cause untold trouble for man, or many other creatures of nature, by its capacity to interfere with myriad cells. It has only one cell, yet it changes its shape, its size, its location, and, without willing it, it changes its progeny. If an amoeba, with a single cell, can do so much, there is room yet for wisdom in man. Man will always pursue; he will always find; and he will always be different tomorrow from what he is today.

Man has been learning slowly for centuries that his own life depends to a great degree on the life of his neighbour. For centuries this process was snail-paced, as when his own sewage was thrown into the street for his neighbour to walk on. Eventually, in self-defence, a community system was devised. Today, every car's exhaust contributes to the irritation of cancer-susceptible cells in each of us.

Once man struggled for the education of only his own offspring. But now one must be aware that the world his child will inherit is only so good, or so safe, as the least-educated child. For the educated man can be destroyed, or mutilated, or enslaved by the man who has no education —no education from home, from school, from community. This man, who has nothing to lose, does not fear to destroy. For 2000 years religious precepts have taught that man must be his brother's keeper, but it remains for science to give example after example of the truth of this early philosophical concept.

There can be no ultimate right, no final understanding, no permanent solutions for the problems of mankind. For change is inherent in the structure of the molecules of which we are composed. This is perhaps the hardest truth, for it allows no rest.

The slogan of yesterday, 'for God and Country', is still the star of today, only the meanings of the words have changed: 'God' is 'the Universe' and 'Country' is 'the Human Race'.

> Respect thyself, lest thy respect be
>   unworthy to give.
> Order thy days, lest thy loved ones
>   be ensnared
>   in thine own confusion.
> Honor the wisdom of thy heart by the quiet
>   of the hour of its borning.

Reflect, for thou art a part
    of all men,
    of all life,
    of all matter
    of all of the stars and
    the voids of eternity.

Then only can the
    beautiful and awesome order
    which is your being
    give to your span of days
    that small new goodness, or
    wisdom, or
    beauty
    which determines tomorrow.

REFERENCE

1. CALVIN, GENEVIEVE J. and CALVIN, MELVIN, Wanted: a star. *Main currents in modern thought* **23,** No. 3, 59 (1967).

APPENDIX

THE POSSIBILITIES OF INTERSTELLAR COMMUNICATION

A RADIO message is reported to have been received from outer space. Recall that the message consists of a series of two symbols and that this series numbers a total of 551 in all. Presumably it is received on a radio receiver in terms of 551 signals, followed by an interval and then the repetition of the same pattern of 551 signals. The problem then becomes one of decoding the simple sequence of 551 binary symbols. Now there are a variety of ways one might go about thinking of a solution. The first one that comes to mind is to think in some sort of binary code. However, a little more subjective consideration of the nature of a message of this kind and its purpose will lead one to quite a different mode of procedure.

After all, there is no certainty that the receiving group or even the sending group have the same method of thought and reasoning. The only factor that one can think of that might have a reasonable possibility of being common to all living organisms, at least in the general context of what we consider to be living, will be the ability to respond to electromagnetic radiation of a variety of wavelengths. Aside from instrumental responses, the individual response can be thought of in terms of the various physical and chemical factors to which any physical-chemical system can respond. There are, of course, electromagnetic radiations of a variety of wavelengths, particularly the most abundant ones within any kind of solar system that could support life of the sort that we can conceive of. The principal physical factor which is common to all such environments which require a specific temperature range would be the electromagnetic spectrum of the centre of that solar system, its sun. This electromagnetic spectrum, in order to have the kind of temperature distribution on the planetary system from which the organism might develop, would have to have a major component in what we consider the visible region, between the wavelengths of roughly 350 or 400 nanometres and 700 nanometres. Such a factor would necessarily produce an organism sensitive to these particular wavelengths and furthermore, the amount of information that electromagnetic energy of this wavelength can transmit is increased when it is transmitted in the form of reflected image rather than in any other simple coded form.

Furthermore, the reflected image would not require any commonality in communication mode other than the use of a set of wavelengths which would produce a reflected image. We thus arrive at the notion that this linear array of 551 symbols might in some way transmit an image. With this conclusion, we are immediately induced to think of the way in which a television system produces an image. As you know, it produces that image by a series of integral intensities of a particular wavelength spread in a raster over two dimensions. This immediately suggests only two intensities, 0 and unity, and these intensity units are to be distributed in a two-dimensional array of some sort. How can we determine what the two-dimensional array is? The number 551 is the product of two prime numbers, 19 and 29, and immediately this becomes apparent, then the size and shape of the raster is suggested. It is either a raster of 19 horizontal by 29 vertical units or the other way round.

The figure reproduced below is the correct decoding of the message; the ones are represented as dark space, the zeros as light spaces, and the 551 characters have been arranged into an array of 29 groups of 19 characters.[1]

An explanation of the rationale of the message and its contents, prepared by Frank Drake is as follows:[1]

The first step in the solution of this message is to determine, if possible, the number of dimensions in which the message is written. If one-dimensional, it will be similar to an ordinary telegram; if two-dimensional, it will be similar to a conventional TV picture, although other than Cartesian coordinates might be employed, etc. We would not expect the number of dimensions to be large, simply because ease of decipherment calls for few dimensions. To make headway in this, one may see what factors may be divided into 551. This test reveals that 551 is the product of only two factors, 19 and 29, both prime numbers, of course. This is a good indication that the message is two-dimensional. Trial and error with Cartesian coordinates shows that breaking the message into 29 groups of 19 characters and arranging these in a conventional TV raster gives a clear-cut picture, which is obviously the correct decipherment of the message.

The interpretation of the picture is as follows:

(a) The figure of the manlike creature at the bottom of the picture is obviously a drawing of the being sending the message. We see that it resembles a primate, with a heavier abdomen than we have, and that it carries its legs more widespread than we do. Its head is also more pointed than ours (or else it has a single antenna). One may speculate from this physiognomy that the gravitational acceleration is greater on the home planet of this creature than it is on earth.

(b) The large square in the upper left-hand corner, accompanied by nine smaller objects strung along the left-hand margin, is a sketch of the planetary system of the creature. We see that there are four small planets, a larger planet, two large planets, another intermediate planet, and one last small planet. The system thus resembles our own in basic morphology.

(c) The two groups in the upper right-hand corner may be recognized as schematic drawings of the carbon and oxygen atoms. We deduce from this that the creature's biochemistry is based on the carbon atom, as ours is, and that the oxidizer used in its chemistry is oxygen, also as with terrestrial animals.

(d) A key group of symbols are those just occurring to the right of the four minor planets and the fifth planet. Inspection of these symbols shows that they are simplified a modified binary representation of 1, 2, 3, 4, 5, written in sequence alongside the first five planets. The modification made to the basic binary numbers is the addition to the ends of the numbers of parity bits, where necessary, so that the number of 1's in every binary number is odd. This is similar to computer practice on earth. It is apparently not used here as a check on transmissions, but rather to designate a symbol as a number. In future communications, symbols will certainly

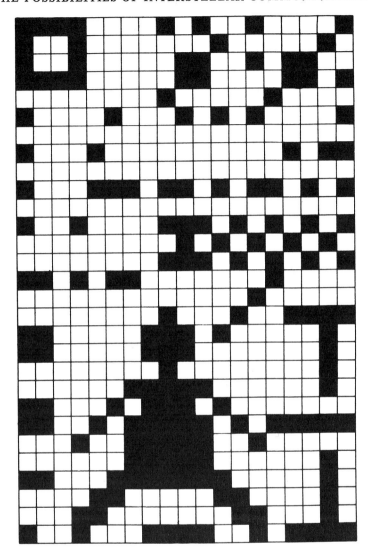

also be used for words of language. We may deduce from the creature's careful setting down of the binary number system that he will use this, with parity bits, for numbers henceforth. It follows that we may expect words of language to have even numbers of 1's. In this way, the creature has established a number system and has enabled us to recognize words of language. (The more detailed interpretation of this message, with more speculation, may be found in reference (1).)

REFERENCE

1. SHKLOVSKII, I. S., and SAGAN, CARL, *Intelligent life in the universe*, pp. 423–7 and references cited therein. Holden-Day, San Francisco (1966).

GENERAL BIBLIOGRAPHY

PART I. THE VIEW FROM THE PRESENT TOWARDS THE PAST

1. ABELSON, P. H. (editor), *Researches in organic geochemistry*, Vol. II. Wiley, New York (1967).
2. BREGER, I. A. (editor), *Organic geochemistry*. Macmillan, New York (1963).
3. COLOMBO, U., and HOBSON, G. D. (editors), *Advances in organic geochemistry*. Pergamon Press, London (1963).
4. DEGENS, E. T., *Geochemistry of sediments*. Prentice-Hall, Englewood Cliffs, New Jersey (1965).
5. HOBSON, G. D., and LOUIS, M. C. (editors), *Advances in organic geochemistry, 1964*. Pergamon Press, London (1966).
6. NAGY, B., and COLOMBO, U. (editors), *Fundamental aspects of petroleum geochemistry*. Elsevier, Amsterdam (1965).
7. RUTTEN, M. G., *The geological aspects of the origin of life on earth*. Elsevier, Amsterdam (1962).

PART II. THE VIEW FROM THE PAST TOWARDS THE PRESENT

1. BERNAL, J. D., *The physical basis of life*, Routledge & Kegan Paul, London (1951).
2. —— *The origin of life*. World Publishing Co., New York (1967).
3. CALVIN, M., *Chemical evolution* (Condon lectures). University of Oregon Press, Eugene (1961).
4. CRICK, F. H. C., *Of molecules and men*. University of Washington Press (1966).
5. EHRENSVARD, G., *Life: origin and development*. University of Chicago Press (1962).
6. FOX, S. W. (editor), *The origins of prebiological systems and of their molecular matrices*. Academic Press, New York (1965).
7. HALDANE, J. B. S., *Origin of life, New Biology*, No. 16. Penguin Books, London (1954).
8. JUKES, THOMAS E., *Molecules and evolution*. Columbia University Press, New York (1966).
9. KENYON, DEAN H., and STEINMAN, GARY, *Biochemical predestination*. McGraw-Hill, New York, in press.
10. KEOSIAN, J., *The origin of life*, 2nd edn. Reinhold, New York (1968).
11. OPARIN, A. I., *The origin of life* (3rd Eng. edition, trans. by Ann Synge). Oliver & Boyd, London (1957). See also earlier edition trans. by S. Margulis. Macmillan, New York (1938).
12. OPARIN, A. I. (editor), *The origin of life on the earth* (proceedings of international symposium sponsored by International Union of Biochemistry, Moscow, U.S.S.R., August 1957). Pergamon Press, London (1959).
13. SHAPLEY, HARLOW, *Of stars and men*. Beacon Press, Boston, Mass. (1959).
14. SIMPSON, GEORGE G., *The meaning of evolution*: a study of the history of life and of its significance for man, revised edition. Yale University Press, New Haven (1967).
15. WOOLDRIDGE, DEAN E., *The machinery of life*. McGraw-Hill, New York (1966).

PART III. THE VIEW FROM THE PRESENT TOWARDS THE FUTURE

1. MAMIKUNIAN, G., and BRIGGS, M. H., *Current aspects of exobiology*. Pergamon Press, London (1965).
2. SHLOVSKII, I. S., and SAGAN, CARL, *Intelligent life in the universe*. Holden-Day, San Francisco (1966).
3. SULLIVAN, WALTER, *We are not alone*. McGraw-Hill, New York (1964).
4. PITTENDRIGH, C. S., VISHNIAC, W., and PEARMAN, J. P. T. (editors), *Biology and the exploration of Mars*. NAS–NRC Publication No. 1296. Washington, D.C. (1966).
5. SHNEOUR, E. A., and OTTESEN, E. A. (editors), Extraterrestrial life: an anthology and bibliography (supplement to *Biology and the exploration of Mars*). NAS–NRC Publication No. 1296A. Washington, D.C. (1966).
6. *NASA 1965 Summer Conference on Lunar Exploration and Science, Falmouth, Massachusetts, July 1965*. NASA Publication SP-68. Scientific and Technical Information Division, NASA, Washington, D.C. (1965).
7. HESS, W. N. (editor). 1967 Summer Study of Lunar Science and Exploration, University of California, Santa Cruz, July–August, 1967. *NASA Publication SP*-157. Scientific and Technical Information Division, NASA, Washington, D.C. (1967).
8. *Planetary exploration 1968–1975. Report of study by the Space Science Board, H. H. Hess, chairman*. National Academy of Sciences–National Research Council (1968).

AUTHOR INDEX

SUBJECT INDEX

DATE DUE

MAR 31 1970		
2-22-73		
11/12/63		
GAYLORD		PRINTED IN U..S.A.